ELECTRONICS RELIABILITY AND MEASUREMENT TECHNOLOGY

ELECTRONICS RELIABILITY

AND

MEASUREMENT TECHNOLOGY

Nondestructive Evaluation

Edited by

Joseph S. Heyman

NASA Langley Research Center
Hampton, Virginia

NOYES DATA CORPORATION
Park Ridge, New Jersey, U.S.A.

Published in the United States of America by
Noyes Data Corporation
Mill Road, Park Ridge, New Jersey 07656

10 9 8 7 6 5 4 3 2 1

Library of Congress Cataloging-in-Publication Data

Electronics reliability and measurement technology.

    "The Electronics Reliability and Measurement
Technology Workshop was held June 3 through 5, 1986
at NASA Langley Research Center"--
    Bibliography: p.
    Includes index.
    1. Integrated circuits--Testing--Congresses.
2. Integrated circuits--Reliability--Congresses.
3. Non-destructive testing--Congresses. I. Heyman,
Joseph S. II. Electronics Reliability and Measurement
Technology Workshop (1986 : NASA Langley Research
Center)
TK7874.E486   1988   621.381'028'7   88-25395
ISBN 0-8155-1171-X

Foreword

This book examines electronics reliability and measurement technology. It identifies advances in measurement science and technology for nondestructive evaluation, and it details common measurement trouble spots. The book is based on a workshop held at NASA Langley Research Center in June 1986.

The objectives of the research described in the book are to improve reliability, yield, performance, and speed, and to reduce failure rates, delivery times, and costs; i.e., to achieve higher quality and to enhance productivity. If these objectives cannot be attained, the U.S. electronics industries face a crisis which may threaten their very existence.

For the U.S. electronics industries to remain in the forefront of the field, they must be able to make quantitative measurements in rapid, nondestructive fashion. It is essential to know which physical, chemical and electrical properties are critical, and to what resolution they must be determined. Various aspects of the subject are discussed in the book, including wafers, parts, and assemblies; and research results are reported. Key problem areas are noted along with recommendations for future research and communication in this most competitive field.

The information in the book is from *Electronics Reliability and Measurement Technology,* edited by Joseph S. Heyman, NASA Langley Research Center, for the National Aeronautics and Space Administration, August 1987. It is a summary of a workshop held in June 1986, at NASA Langley Research Center, and sponsored by NASA, the U.S. Air Force, the National Security Industrial Association, and the Aerospace Industry Association.

The table of contents is organized in such a way as to serve as a subject index and provides easy access to the information contained in the book.

Preface

The Electronics Reliability and Measurement Technology Workshop was held in June 1986 at NASA Langley Research Center and was presented with the cooperation of the National Security Industrial Association (NSIA), NASA, the United States Air Force (USAF), and the Aerospace Industry Association (AIA). This ambitious meeting was held to examine the U.S. electronics industry with particular focus on reliability and state-of-the-art technology. The goals of the workshop were:

> To provide a forum for identifying advances in measurement science and technology for nondestructive evaluation in electronics, to bring common measurement problems to a critical focus, and to encourage improved technology to achieve higher quality, enhance productivity, and foster interaction among government, industry, and university people.

In addition, the workshop had specific objectives. Obvious objectives included to improve reliability, yield, performance, speed, and density and to reduce failure rates, delivery time, and cost. A less obvious objective was to learn how to make these improvements faster than world competition or be forced to accept second-class market and security positions. The mechanisms to achieve these objectives included improving nondestructive evaluation measurement science for the electronics industry, consolidating generic research, and providing a mechanism for better collaboration among industry, government, and university laboratories. A major question involved what roles NASA, USAF, universities, and industry should have in improving the measurement science/nondestructive evaluation technology essential to maintain world leadership in aerospace electronics.

The findings of the workshop addressed various aspects of the industry from wafers to parts to assemblies. Even though there was great diversity brought to the meeting by the approximately 75 government, industry, and university participants, a strong theme was apparent: The U.S. electronics industries are facing a crisis that may threaten their existence, and this crisis is a result of complex issues.

For the U.S. electronics industries to advance, it is not sufficient to solve specific problems; we should develop, literally, an acceleration in the learning curve for electronic and materials properties and for the control of manufacturing processes. The speed of our learning depends on our ability to know what we have before us. Micromeasurements show us that our beautiful polished surfaces can be rough terrains strewn with "boulders." Physics models show us that uniform electrical fields can be complex and dependent on defects.

Quantitative measurements are the only way we can identify where we are. For these reasons, it is important to know which physical, chemical, and electrical properties are critical and to what resolution they must be determined. We must build the instruments that will give us those answers and make those instruments available to the U.S. electronics industries to solve their generic problems.

Other major research areas have been developed in a cooperative fashion. High-energy research groups utilize national labs; industries requiring technical standards use the National Bureau of Standards (NBS) labs; aerospace research groups utilize NASA and Air Force labs. We should work to identify mechanisms for collaboration in electronic nondestructive evaluation (ENDE) measurement science to improve performance, yield, and reliability of electronics while remaining competitive in the world market.

In the Executive Summary of Findings and Recommendations, key problem areas that were singled out for attention are identified and the action items that will work toward their resolution are recommended. Even though the conference was highly rated, it had one flaw that amplified the perceived problem: a small attendance by the manufacturers of integrated circuit (IC) devices. Conversations with the manufacturers that attended the workshop pointed to the problem of curtailed travel in the electronics industries brought on by their present economic downturn.

It is the hope of the conference planners and those in attendance that this effort will result in real changes. To be effective, however, the problem must become a challenge of high-level managers who must look not only at technical problems but also at cultural problems. Innovative and novel applications must evolve from within the industries themselves with intercorporation cooperation becoming the norm. University, industrial consortia, and government laboratories can provide mechanisms to develop an environment suitable for solutions to generic problems. A long-range commitment to such activities is necessary if we are to achieve our former excellence in electronics.

The workshop was possible through the dedicated and hard work of many, with special thanks to the steering committee, the workshop chairmen, and Ms. Pat Gates who helped run the workshop.

Joseph A. Heyman

NOTICE

Contents and Subject Index

NONDESTRUCTIVE SEM FOR SURFACE AND SUBSURFACE WAFER IMAGING......17

Roy H. Propst, C. Robert Bagnell, Edward I. Cole Jr., Brian G. Davies, Frank A.
DiBianca, Darryl G. Johnson, William V. Oxford, and Craig A. Smith

SURFACE INSPECTION—RESEARCH AND DEVELOPMENT......34

J.S. Batchelder

SENSORS DEVELOPED FOR IN-PROCESS THERMAL SENSING AND IMAGING......37

I.H. Choi and K.D. Wise

WAFER LEVEL RELIABILITY FOR HIGH-PERFORMANCE VLSI DESIGN......42

Bryan J. Root and James D. Seefeldt

EXECUTIVE SUMMARY OF FINDINGS AND RECOMMENDATIONS

I. GLOBAL INDUSTRY ISSUES

Conference participants declare a state of emergency in the U.S. microcircuit industry. The demise of the U.S. industry from a leadership position is imminent. Technological accomplishments fostered by favorable sociopolitical climates in Pacific Rim countries are thrusting them ahead of U.S. semiconductor technology. New avenues of innovation, cooperation, and dedication to purpose must be rapidly opened if the threat is to be turned aside. It must be recognized by all interests that any weakening in the technological strength of the U.S. semiconductor industry will result not only in a reduction of the standard of living in this country but also in a severely weakened military posture.

A. <u>Awareness</u> - It is believed that Chief Executive Officer (CEO) and senior-level operating management personnel of semiconductor manufacturing companies, while acutely aware of the foreign threat to their own companies, may not be adequately aware of the scope of the threat to the future viability of the U.S. industry. Where the awareness does exist, it appears that inadequate thought has been given to ways of dealing with it.

<u>Recommendation</u> - The Department of Defense should sponsor a top-level briefing of senior-level and CEO personnel from the top 25 U.S. semiconductor manufacturing companies in order to communicate the gravity of the situation and to get management involvement in and support for corrective actions, some of which are recommended herein.

B. <u>Fragmentation of Resources</u> - There is a multiplicity of integrated circuit (IC) manufacturers, users, and public and private agencies working throughout the IC industry. There is no commonality of purpose or resolve. Without united efforts, there is little hope of countering threats to industry status.

<u>Recommendation</u> - NASA should champion a consolidation of technical efforts among manufacturers, users, and public and private laboratories. In particular, the generic measurement sciences research activities that are conducted at all electronics companies could be consolidated, thus eliminating duplication, freeing resources to attack problems with finesse, and opening the door for industry to concentrate on proprietary research and development that improves their market.

<u>Recommendation</u> - NASA (with the cooperation of other agencies) should sponsor an annual national conference to promote consolidated efforts and to plan and execute strategies.

C. <u>Improved Measurement Science and Technology</u> - In many areas of semiconductor and electronics parts manufacturing and assembly, faulty elements are not found before considerable work has gone into the item. Often an intermittent process/assembly error is not caught before many devices are fabricated. Great improvements in yield and reliability are expected if measurement sciences are advanced, made more quantitative, and brought to the fabrication line as a feedback for control. One cannot manage what one cannot measure—and the foundation for accelerated industrial learning is quantitative information.

<u>Recommendation</u> - NASA (with the cooperation of other Agencies) should develop measurement science technologies for improved electronics reliability concentrating on the 2-to 5-year problems, industry should focus on the near-term problems, and the

1

universities and consortiums, such as Semiconductor Research Corporation (SRC) should address the long-range problems.

   D. Inadequate Information and Data Exchange - There is abundant technical/ research activity occurring in both the public and private sectors; however, exchanges/sharing of findings occur haphazardly.

   Recommendation - NASA and other public organizations should spearhead formation and operation of focused task groups. These task groups should operate in the same spirit as the now ongoing Fine Line Conductor Task Force, sponsored by the Rowe Air Development Center (RADC). Subjects for these task groups should include but not be limited to wafer contamination, sub-micrometer geometry wafer inspection, and others to be defined.

   E. Speed of Technology Transfer - Time is of the essence to sustain industry viability. The committee recognizes that substantial know-how and equipment are developed in academia. Such capabilities would be tremendous tools in the hands of industry for the furtherance of technology, but transfers from university to the semiconductor equipment marketplace occur seldom and slowly.

   Recommendation - The SRC should establish an office to promote rapid technology transfer of manufacturing and metrology equipment developed at universities to the semiconductor equipment marketplace. NASA's Technology Utilization Office may offer a good model.

   F. Manufacturing Science Education - Competitiveness in manufacturing is key to industry survival. No degree programs in IC manufacturing science exist in the U.S. academic system, hence there is no pool of trained graduates from which to staff IC factories. Up to now, manufacturing science is learned on the job.

   Recommendation - The SRC should greatly increase the sense of urgency with which it is collaborating with universities to establish manufacturing science curricula and should establish an accelerated milestone schedule to accomplish this task.

   G. Statistical Process Control - The principles taught by Deming[1] (in which variability of process attributes is controlled) were key items in the enhancement of quality in Japanese industries targeted for improvement. These principles insure process consistency, which in turn insures cost contaminant through scrap reduction, high product yield, and improved prospects for high product reliability.

   Recommendation - While implementation specifics are not obvious, the committee urges the fastest possible promulgation throughout IC manufacturers and users of statistical process control principles in manufacturing science and specifications writing.

   H. Technical Investigations by NBS - The National Bureau of Standards (NBS) is a noteworthy national technical resource. Its involvement in insuring the future of the U.S. semiconductor industry needs to be greatly expanded.

[1] Deming, William Edwards: Some Theory of Sampling, John Wiley and Sons, 1950.

Recommendation - By means of the bulletins and proceedings of this Conference, the NBS is requested to take positive action to increase involvement of staff and equipment in the investigation of:

a. Standards for semiconductor industry
b. Wafer-level reliability testing
c. Calibration of wafer contaminant analysis equipment
d. Providing an analytical software center
e. Other areas to be defined

II. WAFERS

A. Determining Quality of Complex IC Devices - Large-scale integration (LSI) inspection methods are increasingly inadequate for very large-scale integration (VLSI), ultra large-scale integration (ULSI), and very high-speed integrated circuits (VHSIC). A practical wafer inspection method using holography for defect detection to 0.5μm has been demonstrated by equipment at the University of Dayton. This equipment needs to be "commercialized" and made available to industry as rapidly as possible.

Recommendation - NASA should organize a multidisciplinary funding approach to help in commercialization of significant new process equipment and establish a liaison with the SRC to locate and promote candidate equipment.

B. Particle Measurement and Contamination Control - It is recognized that contamination size factors 10 times smaller than producible geometries must be measured and controlled for viable process and reliability results. By 1990 it appears that it will not be possible to measure this contaminant size factor.

Recommendation - A focused task group on contaminant identification should address the issue of accelerating developments in contamination metrology.

Recommendation - Industry working groups should accelerate progress in more affordable and "cleaner" clean-room technologies.

C. Machine Vision Development and Electron Microscopy Sub-Surface Imaging Development - These programs are active at the Universities of Michigan and North Carolina, respectively, under SRC sponsorship. Both are deemed critical by this committee to progress in wafer characterization.

Recommendation - The SRC should continue program funding with a heightened sense of urgency.

Recommendation - The SRC should sponsor increased technical interchange between its principal investigators (PI) and member companies, primarily via visits of PIs to member companies to help them gain better insight into practical needs and problems facing the industry.

D. Whole Wafer Scanning Electron Microscope (SEM) Inspection - This has become common practice throughout the industry and is being done with inadequate knowledge of potential damage to the products due to irradiation and/or contamination. The committee contends that SEM equipment manufacturers have acted irresponsibly in not characterizing the effects (or even the true capabilities) of whole wafer inspection and properly representing true equipment suitability for this task.

Recommendation — This committee should generate a letter that requests the Joint Electronic Device Engineering Council (JEDEC) task group on the MIL-SPEC method 2018 revision to formally contact all manufacturers of whole wafer SEM equipment, state the problem, and request their immediate attention.

E. Wafer Level Reliability Testing — This has become a practical yield enhancement tool used by many IC manufacturers. However, there needs to be more correlation of actual field failures with existing models in order to close the loop relating quality/yield to part reliability. Users should be more aware of the need to return failed parts for more extensive fault identification through expanded failure analysis efforts.

Recommendation — By means of this Conference Proceedings, the IC user community is encouraged to return all failed parts to their own or to manufacturers' failure analysis (FA) labs for fault identification.

Recommendation — IC manufacturers are similarly urged to structure manufacturers reliability analysis systems to encourage failed part return and fault identification so technology can benefit.

Recommendation — Semiconductor Electronic Manufacturers Institute (SEMI), JEDEC, or a related industrial organization should sponsor the formation of a clearinghouse to share "sanitized" technical failure analysis information with all interests in a non-adversarial climate (this is not possible under the present GIDEP* system). Non-adversarial sharing is a key to achieving technological advancements.

F. Test Chips — In concert with wafer-level reliability concerns, test chips have become common; however, every manufacturer has his own, and there is no testing commonality available for production of generalized yield models. NBS has a well-developed test chip program, especially for electromigration.

Recommendation — NBS should develop generic test chip programs. Test chips and programs should be made available to manufacturers for formulating whatever proprietary applications they may need.

G. Second-Generation Computer-Aided Manufacturing (CAM) Software — The lack of data-crunching power and/or adequate decision-making software limits the ability of (CAM) to provide real-time positive feedback into process control. Software is being developed to overcome this problem, but availability is fragmented and limits industry progress.

Recommendation — SRC or SEMI should form a CAM software user's group for rapid promulgation of CAM software systems.

III. PARTS

A. Data Source for Standard and Nonstandard Electronic Parts — An adequate source of data on standard and nonstandard parts does not exist. Information failures, supplier ratings, results of characterization, and evaluation would be of great value if they could be shared on an industry, government, and university wide basis.

Recommendation — Goddard Space Flight Center, as NASA's lead center for standard parts, should develop a data base and reference system and make such information available to as broad a user group as possible.
*Government/Industry Development Exchange Program (GIDEP).

B. <u>Design for Testability</u> - IC and discrete designs should be based on testability and reliability parameters. Integration of in-package sensors and "IC health" monitors should be considered.

<u>Recommendation</u> - A government/industry group must put testability requirements into specifications and should help develop the internal sensor technology for such problems as residual gas analysis.

C. <u>Reliability of GaAs IC's</u> - GaAs IC's are being placed into service without a background of reliability evaluation concepts.

<u>Recommendation</u> - An industry/government group should address these requirements. Since the Defense Advanced Research Projects Agency (DARPA) funded the technology development, they would be a logical funding source to provide a reliability evaluation study.

D. <u>Particle Contamination</u> - Particles are a reliability problem in many electronic parts.

<u>Recommendation</u> - Cleaning and packaging procedures should be improved, and the feasibility of testers that can differentiate between conducting and nonconducting particles should be investigated.

E. <u>Pre-Cap Visual Inspection</u> - Visual inspection, which is labor intensive and subject to human error, is not adequate for LSI and VLSI.

<u>Recommendation</u> - Design an automated system to inspect chips, parts, and assemblies.

F. <u>Failure Analysis</u> - The current failure analysis technology is inadequate for VHSIC and VLSI.

<u>Recommendation</u> - Improve metrology and basic measurement sciences for failure analysis.

IV. ASSEMBLIES

A. <u>Measurement Science Quality Assurance (QA)</u> - The field of measurement science and QA should be given higher visibility, should attract more students, and should address generic problems.

<u>Recommendation</u> - Form a task force to raise quality/measurement issues with ONR*, VHSIC, NSF+, NBS, NASA, and improve funding to SRC, Microelectronic Manufacturers Council (MMC), and review programs with the Department of Defense (DOD).

B. <u>Process Control</u> - More effective methods for process control are necessary to improve yield and reliability.

<u>Recommendation</u> - Artificial intelligence (AI) and adaptive process control must be developed and brought into industry. Current activities at NBS should be made aware of new nondestructive evaluation (NDE) and measurement technologies.

*Office of Naval Research (ONR); +National Science Foundation (NSF).

C. Adequate Trade-Offs for Advanced Technology - Use of immature nonstandard parts causes failures, cost growth, and high part count (cost) in projects in NASA and DOD.

Recommendation - Develop project risk assessment of standard versus nonstandard parts use and compatibility with mission needs.

D. Communication and Technology Exchange - There are many redundant efforts in government, industry, and university circles. A stronger collaboration should be established to achieve a "work together" ethic on generic needs.

Recommendation - Develop a task force to define a mechanism that will result in pooling resources for generic problem solution while still encouraging proprietary efforts for product development. This task force should involve NSIA, NASA, DOD, and NBS.

E. Contamination Detection - Electronic devices require a factor of 10 better resolution than the smallest mask device. Thus 0.5 micrometer electronic devices require 0.05 micrometer measurement technologies. In addition to particles, it is necessary to detect chemical contamination and surface films.

Recommendation - A national laboratory should develop a focused program to help solve this problem, which will limit the future of electronics.

F. Bond Inspection - No complete solution exists for bond inspection that is proven and noninjurious to IC's.

Recommendation - Candidate techniques such as X-ray, acoustic, infrared (IR) inspection and others should be developed and tested. A national laboratory should take responsibility for the evaluation of existing commercial equipment. Additional research should be encouraged in developing novel technologies.

MEASUREMENT SCIENCE AND MANUFACTURING SCIENCE RESEARCH

D. Howard Phillips

Semiconductor Research Corporation
Research Triangle Park, North Carolina

ABSTRACT

The SRC* was established to enhance the competitiveness of the U.S. semiconductor industry through the support of university research and education. Emphasis is being placed on (1) creating and maintaining a generic research base in integrated circuit technologies in the U.S. university community; (2) insuring a continuing supply of highly qualified students (and the faculty required to educate them) to support the growth and innovation of the industry; and (3) broadening the U.S. university base of microelectronic research and education through establishment of centers of excellence, seeding new efforts, and developing new curricula.

The research program of the SRC is managed as three overlapping areas: Manufacturing Sciences, Design Sciences and Microstructure Sciences. A total of 40 universities are participating in the performance of over 200 research tasks.

During the past year, the goals and directions of Manufacturing Sciences research became more clearly focused through the efforts of the Manufacturing Sciences Committee of the SRC Technical Advisory Board (TAB).

The mission of the SRC Manufacturing Research is the quantification, control, and understanding of semiconductor manufacturing processes necessary to achieve a predictable and profitable product output in the competitive environment of the next decade.

The 1994 integrated circuit factory must demonstrate a three-level hierarchy of control: (1) operation control, (2) process control, and (3) process design. Operational control covers process flow execution and inventory control. Unit processes and their control must assure that the equipment yield the anticipated results at the expected particulate contamination level. Particle transport to a wafer surface and capture by that surface depend on many variables such as aerosol velocity and particle size. While gravity is a minor mechanism for submicron aerosol capture by a wafer, it can be the dominant mechanism for capture of larger aerosol particles. Diffusion, interception, and inertial impaction are the classical mechanisms by which filters capture submicron particles; they apply to wafers as well. All the aerosol properties that influence these mechanisms will affect particle deposition on a wafer. In addition, the wafer surface itself may be a significant variable primarily through electrical capture forces. Electrical charges on a surface create electrical capture forces that can make up an important mechanism for the deposition of submicron aerosol particles on that surface. Electrostatic forces will receive special emphasis in upcoming studies of particle transport at the Research Triangle Institute. The use of patterned, electrical biased silicon substrates to deliberately create local areas of potentially enhanced electrical capture is proposed as part of the test matrix. The dependence of particle deposition upon electrical forces will thus be studied with higher spatial resolution than heretofore possible. Appropriate test structures include both metallized patterns on oxides and oxidized silicon surfaces.

*Semiconductor Research Corporation (SRC).

Subsurface imaging methods are important for evaluation of device geometries which are inherently three-dimensional. Accurate depth profiling methods are necessary for failure analysis and process evaluation. SRC research at the University of North Carolina at Chapel Hill has shown that the electron-beam-induced current (EBIC), BSE, and Time-Resolved Capacitive Coupling Voltage Contrast (TRCCVC) imaging modes provide information about buried structures and layer thickness. At submicron dimensions, mechanical probing of devices is nearly impossible and usually destructive when applied to integrated circuit fabrication technology. The electron beam of the scanning electron microscope (SEM) is an alternative to mechanical probes and is a standard tool in the microelectronics industry. Unfortunately, many of the SEM imaging modes can potentially induce damage in radiation-sensitive devices. Low energy (< 1 keV), nondestructive SEM imaging techniques that avoid radiation damage are being developed and modeled in this SRC project. Capacitive coupling voltage contrast (CCVC) is a low-energy method that provides both voltage and depth information of structures buried under passivation. CCVC utilizes the dynamic response of low energy electrons near the surface to changes in voltage or differences in structure depth. Voltage resolutions of 40 mV have been recorded using this technique, with a spatial resolution of less that one micrometer. Depth profiling has been performed on both biased and floating devices. Analysis can be performed during manufacturing, since no bias is required and no mechanical probing is needed. The inclusion of adaptive control will be a result of fully functional computer-integrated manufacturing (CIM) and computer-aided fabrication (CAF). Process design covers the monitoring and feedback functions to assure that the process parameter targets are met and that process-induced defects are kept in control. The SRC/University of Michigan Program in Automated Semiconductor Manufacturing is using reactive ion etching (RIE) as a process vehicle for examining key issues associated with automated sensing, control, and facilities integration. A dual-chamber SEMI-1000 RIE is now installed and is being interfaced with a 10Mbs MAP network in Michigan's new Solid-State Fabrication Facility. This network will be used in a prototype test bed for process automation, offering a guaranteed response time, message prioritization, and a hierarchical facility architecture. Software drivers for this network have now been developed as well as a network interface for the SECS protocol. Companion work in advanced RIE process modeling now allows the calculation of two-dimensional microstructure etch topographies for a variety of etching parameters. These models will facilitate the automatic fine-tuning of etch characteristics and process interpretation via an RIE expert system, also under development. Work on monolithic integrated sensors for pressure, gas flow, and gas analysis will provide additional data for equipment control.

The challenge is to continue to refine the process by which the industry, government, universities, and the SRC staff carry out semiconductor research to increase the effectiveness of this team effort and to provide a more productive response to the need for a continuing flow of new knowledge, creativity, and innovation.

The Members Dollar

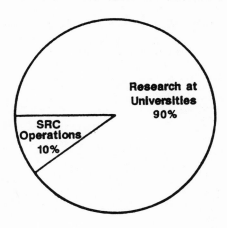

SRC STRUCTURE

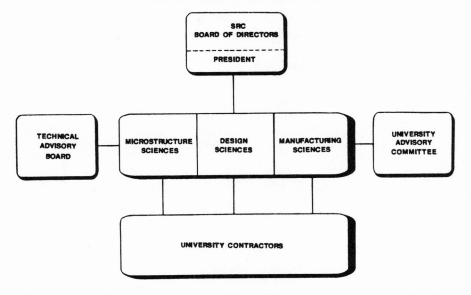

An Industry View of the Future
Direction of Manufacturing Science
(through 1995)
Generated by the TAB Manufacturing Sciences Committee

Industry Contributors

Dr. R. C. Dehmel	Intel (Chairman)
Dr. S. A. Abbas	SRC, IBM
Dr. J. N. Arnold	AT&T Bell Labs
Dr. M. M. Beguwala	Rockwell International
Dr. R. M. Burger	SRC
Dr. B. L. Crowder	IBM
Dr. H. Dixon	Eaton
Dr. S. V. Jaskolski	Eaton
Mr. U. Kaempf	HP Labs
Mr. G. Kern	Monolithic Memories
Dr. W. Lindemann	CDC
Mr. S. A. Martin	Harris Semiconductor
Dr. M. E. Michael	Westinghouse
Dr. D. A. Peterman	Texas Instruments
Dr. D. H. Phillips	SRC
Mr. R. P. Roberge	Union Carbide
Mr. J. L. Saltich	Motorola
Dr. C. Skinner	National Semiconductor
Dr. W. Snow	SEMI Chapter
Dr. P. W. Wallace	SRC
Dr. I. Weissman	Varian
Mr. L. Welliver	Honeywell
Mr. H. Wimpfheimer	IBM

Future Directions of the
Manufacturing Sciences Program

1. Microelectronics CAM/CAF

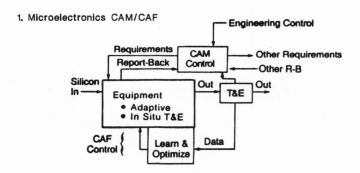

2. Process Inspection and control

   - Measurement/analysis automation
     Machine vision

3. Microelectronics Manufacturing Engineering

   - University – SRC Program

MAJOR TRENDS — SEMICONDUCTORS

- Submicron, MLM, CMOS drives manufacturing
  - 8 Inch Wafers — Now through 1990
  - 10 Inch Wafers — 1990 through 1995

> **THE FUTURE = AUTOMATED MANUFACTURING**

Sources: Genus and SRC

EQUIPMENT IMPLICATIONS

Commodity

- Lowest yielded die cost
- Optimized, dedicated equipment
- High capacity, high utilization
- High operating reliability
- Machine automation, CAM
- Factory automation
- 8 and 10 inch wafers

ASIC

- Shortest manufacturing cycle
- Flexible, dedicated equipment
- Lower capacity, high availability
- High operating and standby rel.
- Machine automation, CAM
- Factory data base automation
- 6 and 8 inch wafers

> **AUTOMATED, ADAPTIVE MANUFACTURING**

Sources: Genus and SRC

SEMICONDUCTOR EQUIPMENT BUSINESS

- 580 companies worldwide (1985)

  - 72% USA
  - 19% Japan
  - 9% Europe

- Top 50 supply 72% of total sales

- Most are 1 product companies

- Average sales = $10.9 M/yr/company

- Highly fragmented industry

Source: VLSI Research Inc., and Genus

Semiconductor Capital Investment Plans
(Unit = $1 Million)

Company	Initial Investment	Percentage Change from Previous Year	Fiscal 1985 Actual Investment (Estimated)	Percentage Revision from Plan	Percentage Change from Previous Year	Planned Investment	Percentage Change from Previous Year
Hitachi	450	0	300	-33.3	-33.7	N/A	Down
Toshiba	600	- 18.9	500	-16.7	-32.4	N/A	Down
Mitsubishi Electric	350	- 0.1	290	-17.1	-17.3	225	-22.4
NEC	700	0	600	-14.3	-14.3	N/A	Down
Fujitsu	500	- 23.7	250	-50.0	-61.8	N/A	Down
Oki Electric	160	- 12.1	125	-21.9	-31.3	N/A	Down
Fuji Electric	65	- 11.0	60	- 7.7	-17.8	35	-41.7
Matsushita	450	- 20.0	300	-33.3	-45.0	200	-33.3
Sony	200	127.9	175	-12.5	86.9	175	0
Sanyo	60	—	61	1.7	—	26.5	-56.6
Tokyo Sanyo	230	46.0	235	2.2	49.2	150	-36.2
Sharp	200	14.3	180	-10.0	2.9	N/A	Level
Total	3,965	- 3.2	3,076	-22.4	-24.9	—	Down

Source: Nihon Kogyo Shimbunsha

WORLDWIDE SALES FORECAST

($ Millions)

	1979	1984	1989	CAGR
Semiconductor ICs[1]	10,500	16,500	37,100	13.5%
Semiconductor Equipment[2]	1,417	6,306	8,082	18.9%
Wafer Processing Equipment[2]	714	3,080	3,798	18.2%
Deposition Equipment[2]	214	662	980	16.4%
CVD Equipment[2]	49	218	326	20.9%

Sources: [1]Montgomery Securities; Jan. 1986 (for years 1980, '85, '90)
         [2]VLSI Research, Inc., and Ganus

1985 SEMICONDUCTOR EQUIPMENT SALES

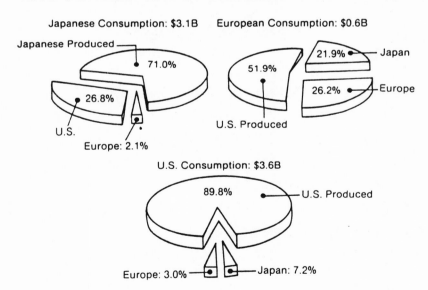

Japanese Consumption: $3.1B   European Consumption: $0.6B

Japanese Produced — 71.0%
26.8%
U.S.
Europe: 2.1%

21.9% — Japan
51.9%
26.2% — Europe
U.S. Produced

U.S. Consumption: $3.6B
89.8% — U.S. Produced
Europe: 3.0%   Japan: 7.2%

Source: VLSI Research, Inc

MAJOR TRENDS — SEMICONDUCTOR EQUIPMENT

- Device requirements drive equipment
- New technologies, materials
  — obsolescence of old
- In situ integration of manufacturing steps
- CAF and CAM = Automated adaptive manufacturing
- Business shakeouts, consolidations
- IC manufacturer, more involved in equipment technology developments
- Joint customer/supplier ownership of performance in operating environment

PROGRAMS

- **University of California, Santa Barbara - GaAs HEMT**
- **Stanford University - Manufacturing**
- **University of Arizona - Packaging**
- **University of Florida - Submicron bipolar**
- **MCNC - Manufacturing**
- **Clemson University - Reliability**
- **RPI - *In situ* processing**
- **University of Illinois, Urbana-Champaign Reliable architectures**
- **MIT - Advanced processing**
- **University of Michigan - Manufacturing automation**

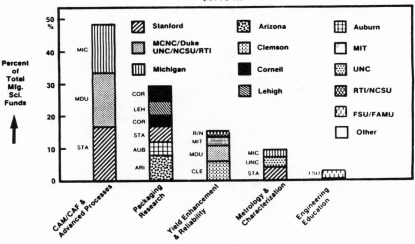

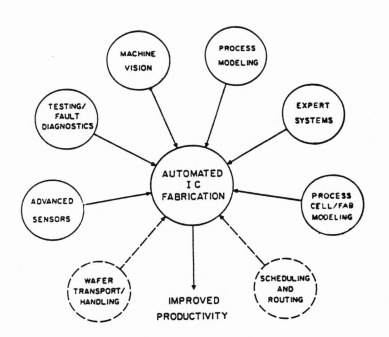

THE SRC/UNIVERSITY OF MICHIGAN

PROGRAM IN AUTOMATED SEMICONDUCTOR MANUFACTURING

Manufacturing Sciences
Significant Accomplishments

1. Stanford   - Microcapillary Cooling Technique
                - Thermal Resistance of 0.04°C/Watt/cm^2

2. MCNC   - Shallow Junctions Demonstrated; 0.15 Micron
                - Germanium Pre-amorphitization,
                  Followed by Boron Implant Results in
                  4x Hole Mobility

3. Michigan   - Developed 32 Element Thermal Imager
                  • Temp. Resolution Less Than 0.1°C
                  • Broad Spectral Range, 0.3 to 30 Microns
                  • Broad Dynamic Range
                - For In-Situ Non-Contact Monitoring

4. Arizona   - Interactive Electrical-Thermal Simulation
                  System
                  • 2-D Capacitance Calculator
                  • 2-D Inductance Calculator

5. Auburn   -   Demonstrated Hybrid Approach to Wafer
                    Scale Integration

6. Publications -   38 approvals into Technical Journals

NONDESTRUCTIVE SEM FOR SURFACE AND SUBSURFACE WAFER IMAGING

Roy H. Propst, C. Robert Bagnell, Edward I. Cole Jr., Brian G. Davies, Frank A. DiBianca,
Darryl G. Johnson, William V. Oxford, and Craig A. Smith

University of North Carolina at Chapel Hill
Chapel Hill, North Carolina

Abstract

The scanning electron microscope (SEM) is considered as a tool for both failure analysis as well as device chacterization. A survey is made of various operational SEM modes and their applicability to image processing methods on semiconductor devices.

I. Introduction

The SEM has become a standard tool for inspection and analysis in the semiconductor industry. Secondary (SE), voltage contrast (VC), electron beam induced current (EBIC), cathodoluminescence (CL) modes, as well as others, including time-resolved capacitive coupling voltage contrast (TRCCVC), provide both static and dynamic methods for investigation. Data from any of the methods are readily digitized into 512 X 512 pixel images. Although each image represents one quarter megabyte of data, image processing can be performed in or near real time.

We have developed processing methods which are particularly suited, but not unique, to semiconductor wafers and devices. The overall goal of our effort has been to show feasibility for various methods which can be used during fabrication as well as after-the-fact analysis.

The SEM is a versatile tool which can provide electrical parameters and topographic information during fabrication. Image processing can be used to enhance both visual and automatic pattern recognition.

II. Image Processing Storage and Retrieval

A. Introduction

Our research is focused on the application of different processing algorithms to Scanning Electron Microscope (SEM) images of integrated circuits. The primary goal is to develop a system to enable failure analysis of integrated circuits using the SEM as a non-destructive analysis tool. The result is the integration of diverse processing methods into a flexible, open-ended and easy-to-use data acquisition and processing system. We have investigated the use of both spatial and temporal domain processing techniques. The former techniques have proved

17

to be the most practical in terms of IC failure analysis purposes. We continue to evaluate new methods in both domains for data acquisition and processing.

Our present hardware comprises two separate systems. The first is a DEC PDP-11/23 minicomputer, coupled with an AED 512 graphics terminal and various A/D and D/A peripherals. This system controls the SEM and the Device Under Test (DUT) and is used for data acquisition and raw data storage. Another PDP-11/23 computer acts as a slave processor to the main 11/23 and is used to control the DUT when complex inputs are required. A block diagram of this system is shown in Figure 1. The second system is a DEC PDP-11/73, coupled with an AED 512 terminal, a video digitizer, and an array processor. This system is used for post-acquisition image processing and software development. The same software package operates on both systems, allowing total functionality with only one system if necessary.

B. Progress

Our work in this area can be divided into three main topics: 1) image acquisition and storage, 2) general processing routines and image data analysis, and 3) information display. The approach taken and the progress made in each of these areas will be discussed separately.

1) Image acquisition and storage

We have established a set of standard techniques for acquiring and storing data from the DUT. The major motivation was a quick and efficient transfer of only the required data from the DUT to the analysis computer. We have developed a method (node scanning) to accomplish this goal. Our findings were outlined in a paper which was published in the December 1985 issue of IEEE Transactions on Reliability.[1]

The basic concept behind node scanning is that a large portion of the area in present-day IC's is occupied by interconnections. Thus, only a small fraction of the area of a die needs to be sampled to determine the voltage present at each of the devices in the circuit. Only this voltage information is necessary in order to determine the operation of the circuit. Further, the raw image data from each of these "areas" (or nodes) can be analyzed and represented in a compact statistical format. Thus, the amount of information required to characterize the entire circuit can be reduced to a standard form which is independent of the actual device geometry. Since the ratio of "active" chip area to "passive" chip area varies widely with circuit design, so does the amount of data reduction using this technique. In this manner, data reduction factors of between 80 and 1300 over standard techniques can be realized. Figure 2 presents this relative data reduction as a function of device density. The major benefit of the node scanning technique is the increased throughput provided by the smaller amount of data which must be processed and analyzed. An additional benefit is the decreased amount of exposure of the DUT to the high-voltage primary electron beam, reducing the probability of damage to the device.

2) General Processing Routines and Image Data Analysis

The largest part of the system consists of a set of general image processing routines. These routines usually operate directly on the image data stored in the AED graphics terminal or on the combination of

AED data and additional data stored in disk files. Our investigation has concentrated on variations of common processing routines which work well with secondary electron (SE) and EBIC image data of IC's. Some of the capabilities we have implemented include image data filtering, addition, subtraction, multiplication, division, contrast enhancement, histogramming, and edge extraction.

One of our main objectives in applying the various processing algorithms to SEM image data is to separate the Voltage Contrast (VC) information from the topographic data. Since the SE signal is dependent on a large number of factors besides local electric fields, the extraction of "pure" voltage information from the SE signal is necessarily complicated. We have developed various image data subtraction techniques optimized for isolation of VC information from secondary images.[2]

Other difficulties associated with making repeatable measurements of voltage levels on the DUT are the non-deterministic characteristics (noise and drift) associated with the image data acquisition system. These factors have been examined in great detail[3] and techniques to minimize their effects have been proposed. In particular, image noise becomes an increasingly large component of the SE signal as the primary beam current is reduced to improve the temporal resolution of the capacitively coupled bound charge decay in a TRCCVC image (see next section). These elements must be dealt with using image processing techniques such as stroboscopic time averaging and filtering. Another aspect of system "drift" can be seen as geometric distortions of the image itself. We have implemented image translation, rotation, and scaling routines which we use to attempt to minimize these distortions. We have investigated various data interpolation methods to minimize the image distortions introduced by the rotation and scaling processes themselves. Edge-detection routines are useful in the detection of geometric shifts and the location of surface features. The performance of several different edge-detection methods on typical SE images has been examined. Image data analysis techniques such as histogramming and thresholded subtraction allow the extraction of data relevant to both the image processing phase as well as the operational analysis phase.

Another area of current interest is in frequency-domain techniques (such as Fourier transform filtering). An investigation of two-dimensional FFT processing for failure analysis and process quality monitoring is currently in progress. Presently, the use of FFT techniques for our image data processing is not desirable because of the computational and memory overhead involved in the FFT process. The majority of the image processing techniques which we use are based on simple 2 x 2 and 3 x 3 convolutions, which are more efficient to perform directly on the image data. We have implemented low-pass and high-pass FFT filtering and the results are nearly indistinguishable from direct convolution filtering. In the future, it may become desirable to perform some processing on the FFT rather than directly on the image data. For example, extremely narrow band-pass filtering is much more computationally efficient when performed using the FFT filtering approach. Thus, we will continue to investigate the applications of FFT image processing techniques.

We are also studying the application of Fourier Transform techniques to wafer and process quality evaluation. The monitoring of process quality directly from the two-dimensional FFT is a complex subject. Unfortunately, comparing the FFT's of two images is not directly analogous to the more conventional approach of comparing two images, since it is not immediately clear how differences in process quality will affect the image, much less the FFT coefficients. In order to understand the relationship between wafer characteristics and the image FFT, the effects of different image acquisition conditions on the FFT have been investigated. The two-dimensional frequency coefficients are dramatically affected by different types of image data and imaging conditions. However, the underlying basic characteristics of the FFT of two images showing the same geometry remain unchanged even though the image contrast and signal-to-noise ratio may change dramatically. The main distinguishing feature of the FFT (the predominant line width) remains unchanged despite differences in the image acquisition conditions.

3)   Image Information Display

The utility of using pseudo-color lookup tables for the display of gray-scale image data has been demonstrated many times. We have developed a facility to produce and modify our own color lookup tables interactively. This has enabled us to develop a set of color tables which is optimized for our purposes. The majority of our data is displayed using a modified Heated Object Spectrum (HOS) lookup table which runs from dark brown at the low end through red and yellow to white at the top end. This allows the use of color to enhance contrast perception but retains the intuitive "order" which is present in most gray-scale tables. In addition, we reserve a small set (typically 32) of pixel values at the top end of the table to allow image features to be marked or outlined and for alphanumeric labeling. These colors are generally "orthogonal" to the colors used in the HOS portion of the color table to allow easy differentiation.

A subroutine has been developed to map the three-dimensional image data function (X,Y and intensity) into a two-dimensional perspective view. This subroutine also includes primitive thresholding and data filtration features. The image intensity variable is represented as the vertical (Z) axis, and the normal X and Y coordinates are mapped into a perspective plane which is at 11 degrees to the CRT plane in the X direction and at 23 degrees inclination in the Y direction. These angles were chosen to enhance the visibility of image contrast features without losing a sense of the overall spatial proportions. Several options are available, allowing the user a choice of "front" or "rear" perspectives, data averaging, scaling, and windowing parameters, for enhancement of the resulting plot.

III.   Time Resolved Capacitive Coupling Voltage Contrast

A.   Introduction

Problems encountered using voltage contrast on passivated devices with a scanning electron microscope (SEM) have been well documented.[4-7] To avoid radiation damage from the primary electrons or the x-rays generated from these energetic electrons, low primary electron beam energies must be used. These electrons penetrate only into the uppermost

portion of the passivation layer.[8] A change in potential on a
subsurface structure will polarize the insulating material between the
structure and the surface. The bound surface charge associated with this
polarization produces a transient in the secondary electron signal from
the device. This signal can be used to generate a dynamic image-
capacitive coupling voltage contrast. We have developed a new technique,
Time Resolved Capacitive Coupling Voltage Contrast (TRCCVC), for
determining the amplitude of this voltage transient. This technique
relates the decay time of the dynamic voltage contrast flash to voltage
transitions on a device. The decay times are also dependent on structure
depth. Our techniques and initial results for depth measurement are also
reported.

B. Signal Generation

A primary electron beam with energy > 100 eV will yield a secondary
electron (SE) energy distribution whose shape is determined by the work
function, Fermi level, and other material parameters.[9] There will be a
net charge accumulation on the device if the primary beam current (I_{pe})
does not equal the loss current (I'). At the beam energies used for
TRCCVC there are two major sources of I': SE and backscattered electrons
(BE). The ratio of BE current to I_{pe}, η, is independent of incident beam
energies greater than about 5 kV. At lower energies small variations
with energy occur. The ratio (δ) of SE current to I_{pe} is dependent upon
the material and the energy of the primary beam.[9] If the BE and SE
currents exceed I_{pe}, as is the case for TRCCVC, a net positive charge
will build up on the surface. This charge prevents lower energy SE's
from escaping the surface and will thus decrease the intensity of the SE
image. An equilibrium surface voltage is reached when there is no net
charge accumulated on the device. When structures deeper than the
maximum beam penetration depth change potential, the material between
them and the surface is polarized, introducing a bound surface charge.
The change in the number of SE's caused by this bound charge is the
TRCCVC signal, which decays back to the equilibrium potential by
permitting more or fewer SE to escape.

C. Data Acquisition and Voltage Results

Operating the SEM at standard TV video scan rates allowed the voltage
contrast data to be videotaped and analyzed on a separate image
processing system. The voltages applied to the DUT were square waves
with variable period and amplitude. All periods were long enough to
allow complete decay of the voltage contrast flash in the secondary
electron image.

After the voltage contrast information had been recorded, decay data
were obtained using a video-rate digitizer (Datacube Inc.). The
digitizer converts the analog composite video signal into a digital array
of 512 x 512 picture elements (pixels); each element has a signal
resolution of eight bits. The resulting system has a maximum sampling
rate of 33.3 msec (video rate of 30 frames per second).

Since the incident electron flux of the SEM primary beam is inversely
proportional to the decay time of a given voltage contrast flash, the

primary electron beam current must be carefully chosen. The transient
decay times must be long relative to the system sampling rate. The
required flux level for adequate time resolution produced a poor
signal-to-noise ratio in the secondary electron image (at the primary
beam energy available on our SEM). To increase the signal-to-noise
ratio, multiple frames of the voltage contrast flash, synchronous in time
with the applied square wave, were averaged. The voltage contrast
amplitude in the averaged secondary electron image was then determined as
a function of time.

For sufficiently large voltage transients, a "saturation" effect of
the dynamic voltage contrast signal occurs. The amplitude of the voltage
contrast flash (and therefore the number of secondary electrons leaving
the device surface) is constant for a time interval after the transition
and before the onset of decay. The saturation parameters depend upon the
SEM operating conditions as well as the work function and electron energy
distribution of the passivation. Higher voltage shifts saturate at the
same intensity but remain saturated for a longer time interval.

The monotonic relationship between saturation time and the amplitude
of the negative voltage shift was used to make a voltage calibration
curve. Using multiple frame averaging, the decay data for different
voltage level transients were plotted. Following Menzel's suggestion[8]
that the voltage contrast flash should decay exponentially, a
least-mean-squares fit of the decay data (intensity-versus-time) to an
exponential curve was calculated. The result was a series of exponential
curves, each representing the decay of a given voltage level. The decay
time required by any given flash to reach a fixed target intensity was
then used to quantify the amplitude of the voltage pulse.

The samples used for voltage measurement were an npn power transistor
and a Schottky diode with approximately 0.6 and 1.5 micrometers of
passivation respectively. Initially voltage shifts from 1 to 5 V were
recorded in 1 V steps. The decay time to the target intensity was then
measured and the results plotted. By taking 1 V steps, the data
acquisition time and SEM drift were both reduced. For higher voltage
resolution, separate measurements were made over smaller ranges:
0.75-2.75V, 0.75-2.0V, 2.0-3.0V, 3.0-4.0V, and 4.0-5.0V (Figures 3 and
4). The SEM magnification was increased at higher voltage ranges to
increase primary electron flux, and thereby decrease decay times. Table
I shows experimental conditions and results. For transitions over the
total 1-5 V range the maximum standard error is 106 mV. However, over 1
volt intervals the error varies from 16 to a maximum of 45 mV. These
resolutions equal or surpass those published by Fujioka[10] using a SEM
at 1.0 kV primary electron beam energy with an electron energy
spectrometer accessory.

Table I. Experimental Conditions and Voltage Resolutions Using the Time
Resolved Capacitive Coupling Voltage Contrast Technique on Two Different
Devices.

Device	Primary Beam Energy (kV)	Sample Area (pixels^2)	Frames Averaged	Voltage Range (V)	Standard Error (mV)
Diode	2.50	64 x 64	14	1-5	58.5
Diode	2.50	64 x 64	10	0.75-2.75	28.5
Transistor	1.25	32 x 64	5	1-5	106.1
Transistor	1.25	32 x 64	5	0.75-2.00	44.8
Transistor	1.25	32 x 64	5	2-3	44.3
Transistor	1.25	32 x 64	5	3-4	34.6
Transistor	1.25	32 x 64	5	4-5	16.3

D. Modeling of Decay Data

In order to predict the shape of the SE decay curves, we must know
the SE energy distribution, surface equilibrium voltage, and incident
electron flux. Gorlich[11] suggests using closed form equations for the
SE energy distribution, Eq. 1,[12] and for δ, Eq. 2[9]. In addition to
these two equations, the time dependent potential on the surface, $V(t)$,
was predicted to decay like that across a capacitor in an RC circuit,
i.e. Eq. 3.[11] The net absorbed current, I_{ae}, will be a function of the
number of secondary electrons with energies greater than that of the
surface potential (which is shown below in Eq. 4). It is obvious that
even with Eq. 1 and 4, $V(t)$ cannot be solved in closed form. An
iterative solution can be formulated assuming $V(t=0)$ is known (Eq. 5).
Using constants for SiO_2[11,12] and typical SEM and device parameters in
the above equations we have constructed the $V(t)$ curves shown in Figure
5.

$$N = \frac{E_{se}}{(E_{se} + A)(E_{se} + F)^y} \qquad \text{Eq. 1.}$$

where: N = number of SE
E_{se} = energy of the SE
A, F, and Y are constants of the material
(A = Fermi level and F = work function)

$$\delta = (Q)^{-0.38} 1.11\ \delta_{max}(1-\exp^{-2.3(Q)^{1.35}}) \qquad \text{Eq. 2.}$$

$$Q = E_{pe}/E_{pe\ max}$$

where: E_{pe} = primary beam energy
         max refers to δ value

$$V(t) = (d/A\mathcal{E}) \int_0^t I_{ae}\, dt \qquad\qquad \text{Eq. 3.}$$

where: $V(t)$ = surface voltage
         $\mathcal{E}$ = permittivity of the material
         A = area scanned
         $I_{ae} = I_{se} + I_{be} - I_{pe}$
         d = structure depth

$$I_{ae} = \left[\delta \frac{\int_0^E N\, dE}{\int_0^{50} N\, dE} + \eta - 1.0 \right] I_{pe} \qquad\qquad \text{Eq. 4.}$$

$$V(t+\Delta t) = V(t) + \Delta t\, (d/A\mathcal{E})\, I_{ae}(V(t)) \qquad\qquad \text{Eq. 5.}$$

Qualitatively, the curves exhibit the same decay recorded earlier (including the "saturation" region). However, there are some problems with incorporating this approach into a quantitative model. First, the SE energy distribution of Eq. 1 does not consider the effects of surface contaminants. Second, the value of V(t=0), for a given voltage change on a device, will be a function of device depth, insulating material, and incident beam energy as well as the amplitude of the voltage change.

To improve on the exponential decay fit to our data, we have measured the integral SE distribution from the peak CCVC intensity after a voltage transition (Figure 6A). This intensity information is offset 3.4 volts such that the peak occurs at 0 volts. These data are then used in Eq. 4 above. The offset indicates an equlibrium surface voltage of 3.4 volts. Therefore, a -1 volt change on the surface would result in CCVC decay from 2.4 to 3.4 volts. The calculated decay curves and normalized data for V(t=0)=0.4, -0.6, -1.6 volts (-3, -4, and -5 volts device voltage changes respectively) are shown in Figure 6B. The -0.6 and -1.6 volt curves agree reasonably well with the measured data, but the 0.4 volt curve decay slower than the data. This discrepancy is probably a result of beam current drift, which is significant on our present SEM at low primary beam energies and beam currents (measured values for these data were 1.25 kV and .011 nA). These experiments and contamination effects will be examined on a new SEM, scheduled to arrive in June '86, more suited to our low accelerating voltage and low current research. Then the effects of drift and model deficiency can be separated.

E. Depth Measurement

Eq. 3 predicts that CCVC decay times should be inversely proportional to device depth (d). The TRCCVC decay data for two different areas on

the same die, each with -5 V transitions, were fitted to an exponential
curve in the same manner as earlier data sets.[13] Two VC decay times
over the same intensity range were then calculated from the exponential
fit of each area. The passivation thickness over these areas was
measured using a Nanometrics film gauge correlated to a Gaertner
ellipsometer. The decay times and depths are 292 and 86 msec, and .549 and
1.580 micrometers, respectively. The ratios are:

$$\Delta t_1/\Delta t_2 = 3.40 \qquad \text{and} \qquad d_2/d_1 = 2.85.$$

From these ratios we see that while the relative depths and decay times
agree with the inverse relation, the predicted and measured ratios differ
by 16%. Part of this difference may be a result of the two areas being
disimilar structures, oxide over metal and oxide over Si.

To eliminate this source of error, TRCCVC data for three different
samples of the same die were examined. As can be seen in Figure 7, the
decay data for these samples varied widely. Later measurements using the
ellipsometer showed only a 2% difference in passivation thickness. We
attribute the difference in signal to surface contaminants on each
device, altering the SE energy distribution and the equilibrium voltage.
Future experiments to eliminate surface contamination by plasma cleaning
will be performed to see if these fluctuations in "identical" devices are
eliminated. The fluctations must be resolved before quantitative depth
and voltage measurements on different devices can be modeled.

F. Conclusions

We are developing a new technique, TRCCVC, for quantitative voltage
and depth measurement of buried structures. The low primary beam
energies used in this technique make it non-destructive, even to
radiation sensitive MOS structures. Initial voltage calibration and
qualitative decay modeling have been sucessful. However, the effects of
surface contamination and SEM drift must be evaluated and/or eliminated
before the overall goal, quantitative prediction of CCVC decay from
SEM/device conditions, is achieved.

IV. Non-destructive Subsurface Imaging of Semiconductors

A. Introduction

The maximum penetration depth (R) of the primary beam-sample
interaction volume in an SEM has been modeled as a function of the
primary beam energy (E) and the atomic number and mass density of the
sample.[14] In a silicon sample the dependence of R on the beam
accelerating voltage may be approximated using the experimental range
$R(E) = 0.022 \, E^{1.65}$, where R is the maximum penetration depth in
micrometers and E is the beam accelerating voltage in kV. The size,
shape, and energy distribution of the beam-sample interaction volume are
also affected by the beam accelerating voltage. We are investigating the
effects of changes in beam energy on images obtained using different SEM
imaging modes in order to evaluate the suitability of the SEM for
non-destructive subsurface imaging of semiconductor devices.

SEM images of semiconductors may be obtained in one of two ways. The products of the elastic and inelastic scattering events occurring within the interaction volume may be measured using an external detector and used to modulate a display, resulting in an image related to the amplitude of a beam-induced emission from the sample. Alternatively, changes in the state of the sample itself induced by the primary beam-sample interactions may be measured and used to produce an image.

In order to compare SEM imaging modes, it is useful to represent each mode by a simplified "image function" model, $I(x,y,E)$, where I represents the intensity of the image at a point (x,y) and E is the primary beam energy. We may think of I as the product of two component functions: I_g, which describes the number of events generated as a function of beam and sample parameters, and I_c, which describes the detection/measurement efficiency as a function of beam energy and penetration.

For example, the transfer of energy from primary beam electrons to weakly bound electrons in the sample results in the creation of low energy (< 50 eV) free electrons that may escape from the sample surface (secondary electrons or SE). The SE emission is measured and used to modulate the intensity of an x-y CRT display, resulting in a "secondary" image. For such an image, I represents the number of SE detected at a given location (x,y) with primary beam voltage E. I_g is proportional to the SE generation efficiency in the material(s) present in the sample and to the size and energy distribution of the interaction volume. The total SE generation efficiency is the sum of the weight fractions of each element present in the interaction volume weighted by the elemental SE generation efficiencies. The beam-sample interaction volume is roughly spherical or pear-shaped; free electrons are generated at all penetration depths r, $0 < r <= R$. The energy distribution within the interaction volume has been modeled as a spherical Gaussian distribution centered in the upper half of the sphere. I_e, which represents the probability that the SE escape the sample and are detected, is proportional to $exp(-r/l_e)$, where l_e is the mean free electron path in the sample. Typical values for l_e are 1 nm for metals and 10 nm for insulators. This model, although approximate, is adequate to allow us to conclude that the SE mode is not well suited for subsurface imaging because of the low yield and rapid attenuation of the subsurface SE emission.

We have developed models similar to the above for imaging modes commonly available on standard SEMs: static voltage contrast, back-scattered electron imaging, TRCCVC (see previous section), cathodoluminescence, EBIC, and x-ray spectrometry. We have isolated TRCCVC, EBIC, and x-ray as the most promising for subsurface imaging, and are currently focusing our efforts on these methods.

B. Experimental Progress to Date

We have investigated the suitability of two SEM imaging modes, EBIC and windowed characteristic x-ray spectrometry, for the detection of subsurface structures. EBIC was used to detect buried P-N junctions in bipolar semiconductor devices. X-ray spectrometry was used to detect silicon layers buried underneath surface metallization.

For EBIC, as for SE, I_g is proportional to the free electron generation efficiency and the size and energy distribution of the interaction volume. Ie is proportional to $exp(-d/l_m)$, where d is the distance to the P-N junction at which the signal may be detected and l_m is the minority carrier diffusion length, which is typically much larger than l_e. Note that electrons do not have to escape the surface in order for a signal to be generated; we may measure EBIC current at P-N junctions buried in the bulk of the device.

The following is an example of experiments performed using EBIC to detect subsurface P-N junctions. The base and collector of a bipolar PNP transistor on a Honeywell 2171 test device were used as inputs to an EBIC detector/amplifier. Images of the EBIC signal were acquired as the primary beam voltage was ramped between 5 and 15 kV. At low beam voltages, an octagonal outline corresponding to the sides of the base-collector junction was visible. As the beam voltage was increased, an EBIC signal was detected from the interior of the octagonal region (Figure 8). This additional signal results from electron-hole pair production at the buried base-collector junction.[15]

We have applied this method to other devices with similar results. Beam penetration through silicon layers as well as surface passivation and metal layers has been observed. We were unable to detect structures at depths greater than approximately 6 micrometers because beam accelerating voltages greater than 30 kV are not available on the SEM used for these experiments. We expect that this method is applicable at greater penetration depths than were attainable using this instrument. We are currently investigating the qualitative potential of this method.

We have performed preliminary experiments using windowed characteristic x-ray spectrometry imaging to demonstrate beam penetration through surface metallization runs. For x-ray spectrometry, I_g is proportional to the elemental concentration within the interaction volume and to the primary beam overvoltage, $E - E_c$, over the energy range examined (E_c is the characteristic x-ray energy). I_e is a decreasing exponential function of distance and the elemental mass absorption coefficients. The use of energy windows to isolate characteristic x-rays greatly increases the sensitivity and resolution of the detection system.

The test device for these experiments was an NPN transistor on a depassivated Honeywell test device. A Kevex energy dispersive x-ray spectrometer interfaced to a multichannel integrator was used to measure the x-ray emissions. Energy windows 110 eV wide centered at 1.49 and 1.74 kV were used to detect the Al and Si K series characteristic x-rays. The SEM beam was scanned in a 256 x 256 raster pattern over the device; x-rays in each window were counted for 20 msec at each point of the image. The x-ray counts were used to modulate the intensity of a digital CRT display. Images were acquired with the SEM primary beam at voltages of 5, 10, 20, and 30 kV. At 5 and 10 kV, the silicon window image showed dark regions where the silicon layer was covered by the surface aluminum. As the beam voltage increased past 10 kV, the signal strength in these regions increased. We believe that this increase is caused by penetration of the primary beam through the metallization into the buried silicon layer.

C. Conclusions

We have evaluated the suitability of several common SEM imaging modes for non-destructive subsurface imaging of semiconductor devices. We are pursuing experimental investigation of three promising methods: TRCCVC, EBIC, and characteristic x-ray spectrometry, in an effort to develop useful tools for efficient, non-destructive semiconductor quality control and fault analysis.

V. References:

[1]W.V. Oxford and R.H. Propst, IEEE Trans. on Reliability, Vol. R-34, No. 5, 410 (1985).
[2]R.H. Propst, C. Bagnell, E. Cole, B. Davies, and W. Oxford, SRC Quarterl Report, March 1984.
[3]F.A. DiBianca, C. Bagnell, E. Cole, D. Johnson, and W. Oxford, Scanning Electron Microscopy, Inc. I, 57 (1986).
[4]O.C. Wells, Appl. Phys. Lett. 14, 5 (1969).
[5]D.M. Taylor, J. Phys. D 11, 2443 (1978).
[6]L. Kotorman, Scanning Electron Microscopy, Inc. IV, 77 (1980).
[7]K. Ura, H. Fujioka, and K. Nakamae, Scanning Electron Microscopy, Inc. III, 1061 (1982).
[8]K.E. Menzel and E. Kubalek, Scanning 5, 103 (1983).
[9]H. Seiler, J. Appl Phys. 54(11), R1 (1983).
[10]H. Fujioka, K. Nakamae, and K. Ura, Scanning Electron Microscopy, Inc. III, 1157 (1983).
[11]S. Gorlich, K.D. Herrmann, and E. Kubalek, Proceeedings of the Microcircuit Engineering 84 Conference, Academic Press, 451 (1985).
[12]M.P. Seah, Surface Science 17, 132 (1969).
[13]E.I. Cole Jr., Appl. Phys. Lett. 48(9), 599 (1986).
[14]J.I. Goldstein et al., Scanning Electron Microscopy and X-Ray Microanalysis, Plenum Press, 53 (1981).
[15]Reimer, Scanning Electron Microscopy, Springer-Verlag, 286 (1985).

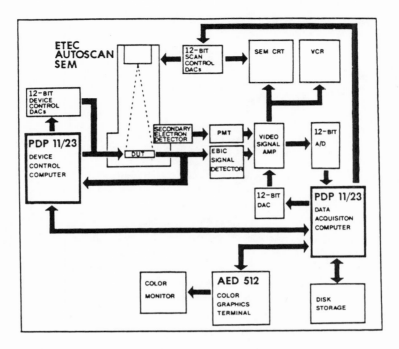

Figure 1.   Block diagram of computer-controlled Scanning Electron
Microscopy image acquisition system.

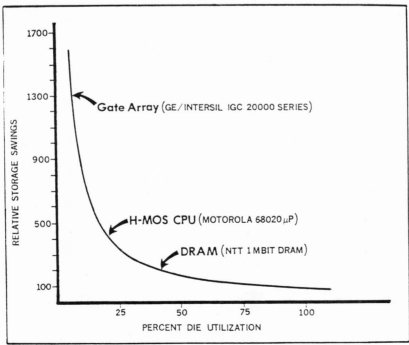

Figure 2.   Data storage comparison between the node scanning method
versus the whole image scanning method at different device
densities.

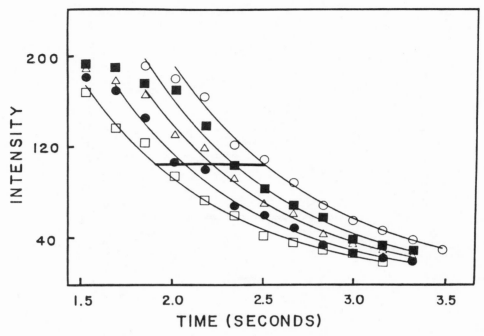

Figure 3.   Measured decay data and best fit exponential curves for -4.0 to -5.0 volts applied transients. The line across the plot is the target intensity value.

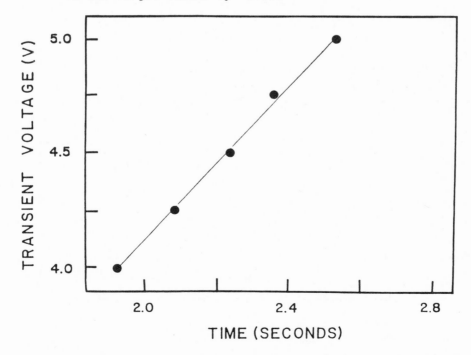

Figure 4.   Time for an applied transient voltage to decay to the target intensity.

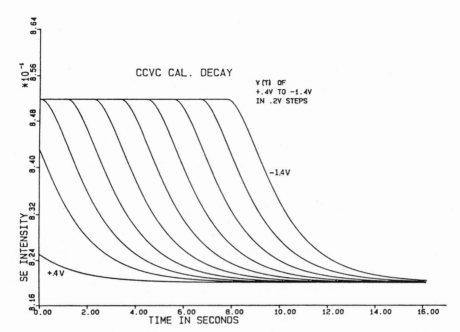

Figure 5. CCVC calculated decay curves from 0.4 to -1.4 volts initial
surface voltage, 0.2 volt steps, using modeled SE energy
distributions.

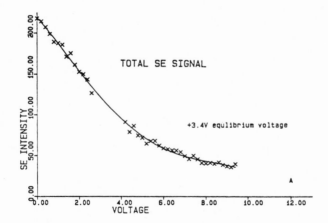

Figure 6A. Measured integral SE response with surface voltage.

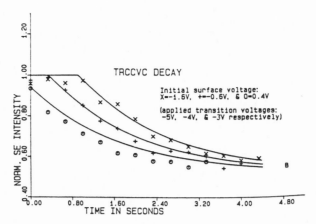

Figure 6B.   Calculated CCVC decay curves from experimental integral SE energy
distribution and decay data for 0.4, -0.6, and -1.6 volt
initial surface voltages.

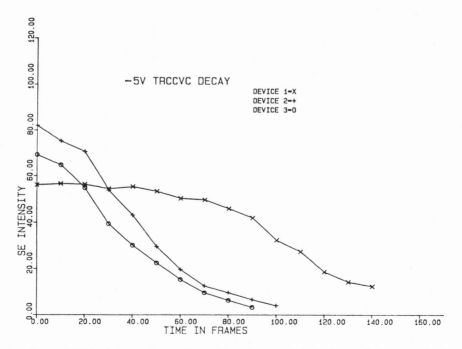

Figure 7.   CCVC decay data for -5 volt applied transitions of the same
device on different dies.

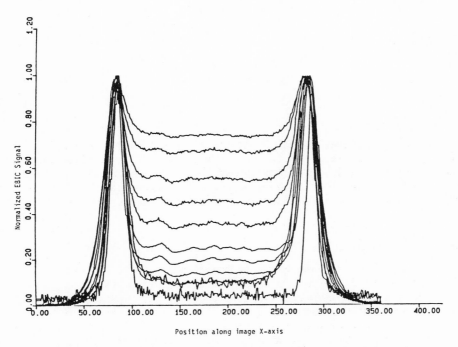

Position along image X-axis

Figure 8.   EBIC signal of base-collector junction at primary beam energies
from 5-15 kV in 1 kV increments.

SURFACE INSPECTION—RESEARCH AND DEVELOPMENT

J.S. Batchelder

IBM Watson Research Lab
Yorktown Heights, New York

Introduction.

Surface inspection techniques are used for process learning, quality verification, and postmortem analysis in manufacturing for a spectrum of disciplines. We will first summarize trends in surface analysis for integrated circuits, high density interconnection boards, and magnetic disks, emphasizing on-line applications as opposed to off-line or development techniques. We will then look more closely at microcontamination detection from both a patterned defect and a particulate inspection point of view.

Trends in surface analysis

While the critical problem of the week on a given manufacturing line will fluctuate dramatically, taken over time there are at least five types of measurements that are key to yield and productivity today, and which will tend to gate future manufacturing capability. One is pattern inspection of integrated circuits and dense interconnection packages. This is done to verify the pattern back to a software data base, and to un-cover process variations and contaminants that manifest themselves as pattern defects. The second is detection of contamination (particulates and asperities) on monitors and product parts. Particles are the single biggest yield killer in the silicon industry, and they will probably retain that distinction for the near future. The third is metrology in two and three dimensions. This would include linewidths, via sizes, trench aspect ratios, and head dimensions and flying heights. This tests processes like deposition, etching, and lithography overlay, exposure, and development. The fourth is film characteristics such as haze, film thickness, roughness, step coverage, and the presence of contaminating films. The fifth is electrical characteristics, which would include dopant concentration, surface passivation, resistivity (including shorts and opens), and oxide integrity.

Looking at surface inspection in the electronics industry in very general terms, we can assert that inspection will become an area of heightened activity for two reasons. To a large degree due to the influence of Japan (and soon Korea), the learning curves of linewidths, yield, defect density, and other critical parameters against time will probably have to be steeper in the future than their historical trends. Inspection is crucial to allow development and manufacturing the ability to learn how to improve, and to learn quickly. A second reason is that both silicon devices and magnetic disk flying heights are moving into the 0.1 to 1 micron size range; a regime where particles are more abundant and they are much harder to remove from surfaces. If particles are a key yield killer today, they will probably be worse tomorrow.

There are several new specific surface inspection problems that deserve attention.

- Integrated circuits are moving towards designs using high aspect ratio trenches. These might be a micron across and more than five microns deep. How should the profile of these trenches be measured without destroying the parts? How can the presence of contamination in the trenches be detected?

- At the termination of Triology's chip manufacturing stage, Amdahl described one of their key problems as eliminating inter-level shorts in the on-chip multi-level wiring. Insulator opens should be expected to be a problem that will require an appropriate measurement for process feedback and control. Contamination is likely to be a major contributor to insulator failure.

- There is a need for an accurate disposable absolute calibration standard for particle size. The features used to simulate particles should be sufficiently similar to actual contamination that a variety of different detection techniques can use the same standard.

New front end detection techniques

Most of the surface inspection work today is done using variations on optical microscopy, ellipsometry, total integrated scattering, and stylus techniques. There are enhancements and alternatives which could be fruitful.

- Parallel optical processes looks very attractive because of its high information processing rate. InSystems has explored holography for mask inspection. Similar techniques might be applied to opaque surfaces.

- Confocal laser microscopy looks promising for accurate non-contact height and lateral metrology. Heterodyned versions give height sensitivity that is better than typical sensitivities for total integrated scatter techniques.

- The use of evanescent wave illumination for particle detection on monitor surfaces could offer considerable improvements in sensitivity, due to the lack of the reflection interference node at the surface.

- Electron beam techniques will become more common. For example, a thermal technique has been demonstrated in which the surface is heated with a rastered electron beam, and particles on the surface are detected by their subsequent infra-red emission.

Tool development

The decreasing defect densities required for the next VLSI generation imply a parallel increase in processing speed for defect and particulate inspection systems. For example, the number of particles per unit area of size greater than some threshold value goes roughly as the inverse area subtended by that particle. If the sensitivity of a particulate inspection technique goes at some constant ratio of the inspection pixel size to the particle size, and the number of allowable particles or defects per chip stays constant to obtain a given yield, then the bandwidth of the inspection technique should increase as the inverse square of the allowable particle size.

The use of monitor surfaces, such as bare silicon, to measure process induced contamination is not the method of choice. Running monitor wafers uses process tool time. Many deposition and etching steps produce surfaces that are so rough that monitor inspection tooling is relatively insensitive. Particle collection rates depend on the features, composition, and chemical treatment of the surface, and are therefore different between monitors and product.

There are allowable local and global variations in the patterns on the surface due to process variables that should not be flagged as defects. These variations can be, for example, larger than the one tenth ground rules limit. Pattern defect inspection systems need to be locally adaptive to ignore anomalies typical of process variation.

The rule will soon be that inspection systems contain the equivalent of a small main frame computer. In order to retain maintainability of the code, reasonable adaptability of the tools, and acceptable software development costs, the same disciplines that have been developed for large military and commercial software systems should be applied to inspection equipment. Optimally, this would allow more rapid up-grades and diversification of the tool's application.

Research topics

Total integrated scatter is not sufficient to detect particles on rough or highly patterned surfaces. We need to understand better the effect of simple geometries present at a surface on the scattered light field. In particular, this would include a step, an ellipsoid, and a line, all with geometries on the order of the wavelength of light, as a function of the incident angles of the illuminating field, and as a function of the roughness and composition of the features and the surrounding surface.

Along the same lines, the inverse scattering problem needs attention; namely, under what conditions can the observed light scatter field be used to infer the structure or composition of the scattering object? What are the critical components of the scattered light field that can be used to determine characteristics of the scattering feature?

There are a host of electronic and optical parallel processing schemes that might be applied to defect detection. Are any of these particularly applicable? Are there alternative algorithms to those in use today that would particularly suite one of these architectures?

SENSORS DEVELOPED FOR IN-PROCESS THERMAL SENSING AND IMAGING*

I.H. Choi and K.D. Wise

*Solid-State Electronics Laboratory
Department of Electrical Engineering and Computer Science
The University of Michigan
Ann Arbor, Michigan*

As large-scale integration requires smaller geometry and larger chip size, fabrication process control becomes more stringent and thus reliable and efficient process evaluation during wafer processing becomes more important. Among various evaluation methods, non-contact non-destructive techniques including thermal and optical methods have been of our particular interest. Temperature profiles can be used to provide a diagnostic tool for in situ process control and to find optimal process conditions. Optical methods can provide material and process characterizing information such as thin-film thickness and quality, carrier concentration, or content of contaminants [1-4].

Although a variety of sensors are available in association with these non-contact methods, infrared detectors are attractive since they can be used for both methods. Most of infrared detectors can be classified into two major categories: photon-type and thermal-type detectors. A thermopile, a type of thermal-type detector, was chosen for our research due to the following reasons. A thermopile can respond from ultraviolet to far-infrared and its sensitivity is almost flat over this region. It can be also operated over a relatively wide range of ambient temperatures as well as at room temperature. Unlike a pyroelectric detector, this detector exhibits a stable response to DC radiant signal and, if appropriately designed, it can show a wide dynamic operating range. These factors are particularly significant for an in situ process or material evaluation system; process-related signals are typically slow-varying and have a broad range of signal levels, and the corresponding sensors may have to be reliably operated at various ambient temperatures.

Compared with conventional thin-film thermopiles, the thermopile we have proposed [5] uses silicon as a substrate material which should have a high thermal conductivity to sink heat effectively. PolySi is used as a major thermoelectric material and gold is used essentially as an interconnecting material to form a thermopile structure. Another important aspect is use of dielectric diaphragm to support the thermopile. Since SiO_2 and Si_3N_4 are relatively high in thermal resistance, a diaphragm formed by a combination of these films can provide an effective thermal isolation between the hot junction and cold junction regions in the thermopile.

Figure 1 shows a photograph of the silicon-based thermopile chip and a schematic cross-sectional structure of the detector. To realize this thermopile, silicon micromachining was cooperated with conventional IC process technology. Since this fabrication process is particularly compatible with standard MOS process, this approach has the potential to contain on-chip signal conditioning and processing circuitry. This aspect becomes particularly significant for detector arrays. It is desirable for an array to have on-chip

* This work was supported in part by the Air Force Office of Scientific Research under Contract No. F49620-C-0089 and by the Semiconductor Research Corporation Contract No. 84-01-045.

readout capability, which will much simplify the whole system and help the system less plagued by noise pickup.

A detector array is useful and demanding to obtain such information as distributions or profiles of certain physical properties. Some examples include temperature profile or uniformity in carrier concentration or film thickness. For such applications, a 32-element linear thermopile array has been designed and successfully fabricated.

Figure 2 shows the linear thermopile array fabricated on a silicon substrate. This array consists of two 16-element linear subarrays and the two subarrays are staggered to each other to improve array spatial resolution. A second subarray collects the radiative signal incident onto the inactive rims between the detector elements in a primary subarray when the whole array is scanned in a direction perpendicular to the linear arrays. A proper time-delay compensation should be determined by the scanning speed and the distance between the two subarrays.

Each thermopile element consists of 40-junction polySi-Au thermocouples and is supported on a rectangular dielectric diaphragm window. The diaphragm window has a dimension of 0.4 mm x 0.8 mm and the diaphragm is 1.3 μm thick. Each thermopile element is separated by a 20μm-thick, heavily-boron-doped silicon rim, which also supports the cold junctions in each thermopile.

Another important aspect of this array is use of on-chip analog multiplexers. There are two 16-channel analog multiplexers and each multiplexer converts 16 parallel outputs from each detector subarray into a serial output. Thus there are only two output nodes required, compared with 32 output nodes without multiplexing. This advantage becomes more significant if the number of detector elements is larger. Each analog multiplexer consists of a NOR-type decoder having 4 control inputs and 16 MOS-type switches.

This array was mounted on a dual-in-line package and was coated with carbon blacks to characterize its optical and electrical performance. Typical detector responsivity in an array ranges from 65 V/W to 80 V/W and their detectivity ranges from 3.6 cm$\sqrt{Hz}$/W to 4.6 cm$\sqrt{Hz}$/W. Nonuniformity in detector responsivity in an array was less than 10 % and interelement cross-talk was less than 3 %. Their response time is distributed between 7 msec to 12 msec when blacks were coated.

A theoretical performance was evaluated for a thermal imaging system using this array while a practical system is under construction. Assuming a typical scanning-mode system which has a Gaussian modulation transfer function (MTF), the assumed system exhibits a noise equivalent temperature difference (NETD) of 0.6 ° C and a minimum resolvable temperature difference (MRTD) of 0.9 ° C at a spatial frequency of 0.2 cycles/mrad. Although this result is reasonably moderate, the imaging performance of such a system could be improved significantly since thermopile detectivity can be further improved and signal pass-band can be further limited in most practical cases associated with process monitoring and evaluation.

On the other hand, this type of array is being considered to be used for laser-based infrared analysis. For example, using a CO_2 laser as a radiation source and measuring infrared absorption or reflection, film thickness or carrier concentration profiles throughout a wafer could be evaluated [6].

As a conclusion, a monolithic thermopile infrared detector array has been developed using conventional MOS technology and micromachining, and this type of array has a potential to be used for an inexpensive non-contact in situ process evaluation system.

ACKNOWLEDGEMENT

The authors would like to express their appreciation to Dr. R. Toth of the Dexter Research Center, Dexter, Michigan for his help in preparing this device, and Mr. F. Schauerte and Mr. J. Biafora of the General Motors Research Laboratories, Warren, Michigan for their assistance in making masks and ion-implanting samples.

REFERENCES

[1]   L. A. Murray and N. Goldsmith, "Nondestructive Determination of Thickness and Perfection of Silica Films," *J. Electrochem. Soc.*, vol. 113, no. 12, pp. 1297-1300, December 1966.

[2]   L. Jastrzeske, J. Lagowski, and H. C. Gatos, "Quantitative Determination of the Carrier Concentration Distribution in Semiconductors by Scanning IR Absorption: Si," *J. Electrochem. Soc.*, vol. 126, no. 2, pp. 260-263, February 1979.

[3]   T. Motooka, T. Warabisako, T. Tokuyama, and T. Watanabe, "Optical Measurement of Carrier Profiles in Silicon," *J. Electrochem. Soc.*, vol. 131, no. 2, pp. 174-179, January 1984.

[4]   P. Stallhofer and D. Huber, "Oxygen and Carbon Measurements on Silicon Slices by the IR Method," *Solid State Technology*, vol. 26, no. 11, pp. 233-237, November 1983.

[5]   I. H. Choi and K. D. Wise, "A Silicon-Thermopile-Based Infrared Sensing Array for use in Automated Manufacturing," *IEEE Tran. Electron Devices*, vol. ED-33, no. 1, pp. 72-79, January 1986.

[6]   G. E. Crook and B. G. Streetman, "Laser-Based Structure Studies of Silicon and Gallium Arsenide," *IEEE Circuits and Devices Magazine*, vol. 2, no. 1, pp. 25-31, January 1986.

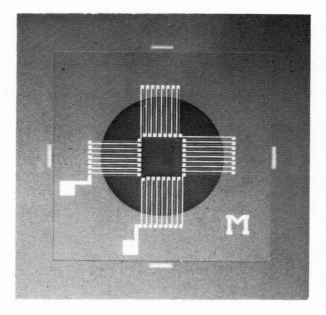

(a) A 32-junction polySi-Au thermopile detector based on a Si substrate. The chip size is 3.6 mm x 3.6 mm.

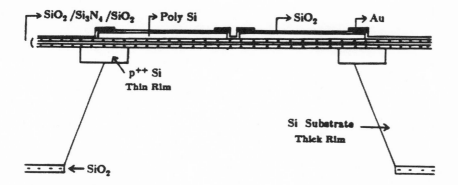

(b) A schematic diagram of the thermopile cross-sectional structure.

Figure 1

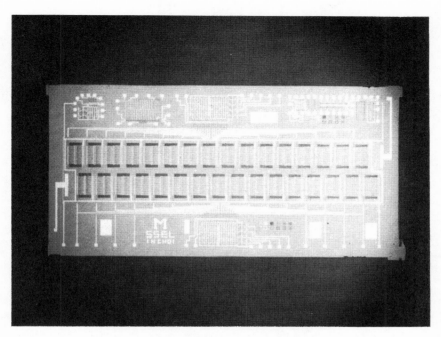

(a) A 32-element thermopile linear array containing two on-chip analog
     multiplexers. The chip size is 11 mm x 5.5 mm.

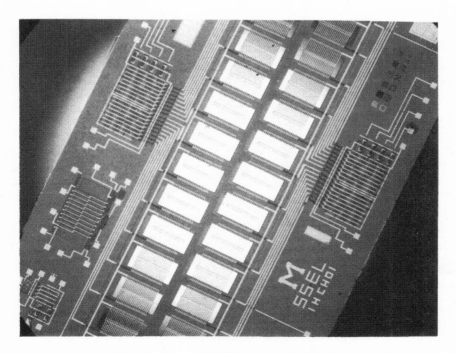

(b) A close-up view of the monolithic array   Each thermopile element is
     supported on a 0.4 mm x 0.8 mm diaphragm window.

Figure 2

WAFER LEVEL RELIABILITY FOR HIGH-PERFORMANCE VLSI DESIGN

Bryan J. Root and James D. Seefeldt

Unisys Corporation
Semiconductor Operations
St. Paul, Minnesota

Abstract:

As VLSI architecture requires higher package density, reliability of these devices is approaching a critical level. Previous processing techniques allowed a large window for varying reliability. However, as scaling and higher current densities push reliability to its limit, tighter control and instant feedback becomes critical. Previously, long–term package level testing that identified 100–year wearout mechanisms was adequate. Misprocessing resulting in slighty reduced wearout did not affect system performance. Due to scaling, however, normal lifetimes are approaching 20 years. Therefore, wafer level tests providing immediate feedback are essential to screen devices susceptible to any premature failure.

This paper describes several test structures developed to monitor reliability at the wafer level. For example, a test structure has been developed to monitor metal integrity in seconds as opposed to weeks or months for conventional testing. Another structure monitors mobile ion contamination at critical steps in the process.

Thus the reliability jeopardy can be assessed during fabrication preventing defective devices from ever being placed in the field. Most importantly, the reliability can be assessed on each wafer as opposed to an occasional sample. Unisys Semiconductor is working on developing this technology.

1. INTRODUCTION

The electronics industry has seen explosive growth in the complexity of semiconductor devices. In only 20 years, computers previously built with discrete devices are now constructed with components containing hundreds of thousands of elements. As these devices become more complex, the reliability of these devices approaches a critical level.

Throughout the history of semiconductor devices, each new generation, even though more complex, exhibited a higher level of reliability. This is due to increased understanding of the physics of semiconductor devices as well as the advent of sophisticated instrumentation. As the industry moves from VLSI architecture to ULSI, the trend of increasing reliability must continue. The primary reason for this is because the lifetime of the devices has contracted to the point that any premature failure results in reduced system performance. Specifically, design rules in the 5μm range and higher normally show device lifetimes that might be in excess of 100 years. If the system life is expected to be a minimum of 20 years, and due to processing irregularities, the semiconductor device has a slightly reduced lifetime, system performance will not be compromised. However, as normal lifetimes of VLSI and ULSI devices approach 20 to 30 years, any premature device failure will degrade system performance.

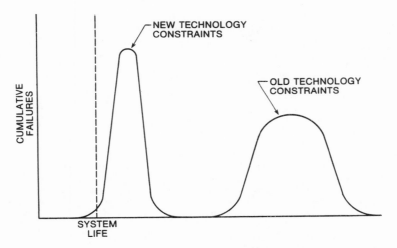

Plot of Lifetimes vs. Cumulative Failures.
While System Life Has Remained the Same, Device Lifetimes Have Contracted.

The conventional methods of reliability screening, that of short–term burn–in to eliminate infant life failures, and long–term life tests at high temperature, will soon become inapplicable for many devices. The reason for this is increasing customization, cost, and shortened lifetimes. As an example, it has already been shown that standard burn–in of DRAMs, [1] while eliminating inferior devices, can substantially shorten the lifetimes of "good" devices to an unacceptable level.

Additionally, applying standard burn–in techniques to a small production run of semicustom devices could result in more devices being used in testing than are delivered to the customer. Thus, the trend is clearly forcing semiconductor manufacturers to adopt wafer level tests for process screening and reliability evaluation.

2. WAFER LEVEL TESTS

In the last four years, most US semiconductor manufacturers have organized research efforts in the area of wafer level testing for reliability, process screening and yield enhancement. This paper describes several wafer level tests under development at Unisys. The emphasis is directed toward ultra high performance bipolar process technology; however, many of these ideas can be applied to "standard" processes or modified to accommodate MOS reliability.

3. WAFER LEVEL ELECTROMIGRATION TESTS

In April of 1985, two papers were presented at the International Reliability Physics Symposium on wafer level tests for electromigration susceptibility. The first, which is called the Standard Wafer–Level Electromigration Acceleration Test or SWEAT is described here. [2] The other test, Breakdown Energy of Metal (BEM), was developed at Intel. [3]

3.1 Important Considerations in the Development of Accelerated Wafer Level Tests

First, the test must be of extremely short duration to achieve adequate throughput in the production line. Second, the test cannot impact the reliability or yield of the normal functional part. Third, the test must correlate to established data for conventional testing.

The SWEAT test makes use of the well–known equation for electromigration: [4]

$$MTF = AJ^{-N} \exp(e_A/kT) \qquad\qquad [1]$$

where: MTF is the median time to fail
A is determined by the metallization
J is the current density through the metal lines
N is the current density factor
e_A is the activation energy of the electromigration
K is Boltzmann's constant
T is temperature in Kelvin

The important variables to note in equation 1 are the current density and temperature. In normal electromigration testing, the current density is often increased to 10X above operating current densities and the device is placed on a hot chuck or in an oven to achieve a failure in a timely manner. This type of testing has been found to be valid where significant Joule heating of the metal strip is not present. This is due to the inability to accurately monitor the temperature. Thus, this type of testing takes days and often weeks to obtain statistically significant data.

To obtain a highly accelerated test for electromigration (i.e., to obtain failures in seconds as opposed to hours), the temperature and the current density must be raised to extremely high levels. This presents a problem in control and monitoring. For precise acceleration factors, the current density and temperature must be known, controlled and updated continuously during the test. The SWEAT test controls the temperature by correlating it to the instantaneous power density, where the power density equation is given in equation 2.

$$Power Density \equiv P = (VI)/V_L \qquad Watts/cm^3 \qquad\qquad [2]$$

where: V is the voltage
I is the current
V_L is the volume of the line under stress

Thus, by monitoring the current through the test device, and controlling the temperature using power density, a constant acceleration factor can be applied. Equation 1 can be modified as:

$$TT_F = A(I/cm^2)^{-N} \exp[e_A/K(PQ - T_0)] \qquad\qquad [3]$$

where the modified temperature expression is given by

$$T = PQ - T_0 \qquad\qquad [4]$$

where P is the power density in equation 2 and Q is a constant slope of power density vs. temperature given in units $((cm^3 - K)/W)$, and T_0 is the ambient temperature. The power density vs. temperature curve is derived by first correlating resistance vs. temperature curves on a variable temperature wafer chuck. Power vs. temperature is then derived by raising the temperature of the line by increasing levels of power (I^2R); no external heat is used.

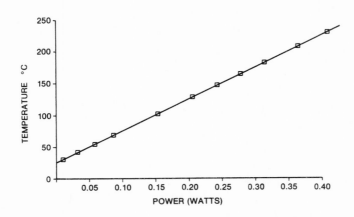

Plot of Temperature vs. Power.
Note in All Cases This Has Been Found to Be a Linear Relationship.

3.2 Implementation of the Test

The implementation of the SWEAT test first requires characterizing what is considered to be minimally acceptable material. This is because the SWEAT test is designed as a simple pass–fail test. In a normal production mode, the benchmark established with the minimal acceptable material is compared with the production material. Thus, if the minimum specification material is set up to last an equivalent time to fail (ETT $_F$) of 15 seconds, then the current and power density in the production material is compared against this. Then equation 3 becomes:

$$\text{ETT}_F \equiv \text{TT}_F = A(I/cm^2)^{-N} \exp\left[e_A/K(PQ - T_0)\right] \qquad\qquad [5]$$

where TT $_F$ of the subject material is increased until it is equivalent to ETT $_F$.

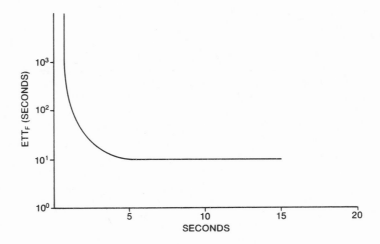

The Calculated TT $_F$ Zeros in on the Target ETT $_F$ During the Test.

3.3 Electromigration Test Structure

It has been known for some time that various physical parameters such as grain size, passivation, mechanical stress, metallization type and topography effect electromigration sensitivity. It is possible to concentrate several stresses in one area to enhance the electromigration characteristics.

High current gradients have been shown to enhance electromigration. Levine, et. al., [5] showed voids forming just inside the negative bond pad where stress from the electron wind changes rapidly. As discussed previously, temperature greatly enhances electromigration where voids tend to grow toward hot spots [6] in metallization. Additionally, due to current sputtering techniques, metal thinning at steps causes current density gradients and heating, thus enhancing electromigration. A test structure was developed to concentrate these stresses improving the precision and resolution of the test. The structure essentially duplicates worst case conditions in a device. Also, real–time observations have shown voids traveling down substantial portions of the lines before enlarging to an open circuit. In the SWEAT structure voids are trapped in the narrow sections, thus enhancing the sensitivity.

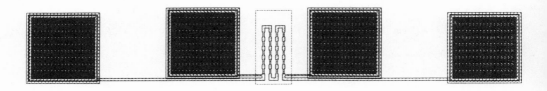

The SWEAT Test Structure.

The test structure shown is a series of wide and narrow regions over topography. The wide areas are designed as thermal sinks to reduce overall temperature and emphasize the electron wind. The narrow regions route over steps which, depending on the structure and worst possible step coverage, can be up over poly or down over diffusion. In current designs there are 20 narrow regions to increase randomness and assure multiple grain boundaries.

A modification of this structure can be used to monitor step coverage on a wafer to wafer basis. It involves using a SWEAT structure without topography adjacent to a SWEAT structure with topography. Comparing the two cancels most other mechanisms, but insolates step coverage as a potential failure.

4. MOBILE ION CONTAMINATION

The phenomenon of mobile ionic charge is a well known yield reduction and reliability concern. Substantial attention has been applied to MOS device degradation due to unstable threshold voltages; however, in bipolar devices contamination by alkali ions such as sodium and potassium or negative ions and trace heavy metal contamination can also cause device failure.

Device degradation in bipolar devices due to mobile ion contamination manifests itself by creating inversion layers. These inversion layers occur in isolation areas creating leakage from device to device. Additionally, with the new ultrahigh performance double–poly emitter bipolar devices, mobile ion charge can cause junction leakages above acceptable levels. Older bipolar devices operate at such high currents that minor leakage does not affect performance. However, newer generation bipolar devices, due to scaling and lower operating currents, result in increased sensitivity to junction leakage.

A rudimentary test structure for wafer level mobile ion contamination has been available for years in the form of CV dots. A new test structure has been designed, however, that avoids any external heating by employing self heating and can be placed into any scribe lane or dropout test die. This test structure is essentially a capacitor structure.

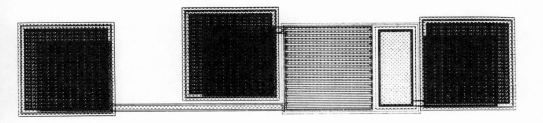

The Mobile Ion Test Structure.
Note the Serpentine Across the Top of the Capacitor Plate.

The lower plate is formed by a silicon epitaxy mesa with an oxide layer cover. The upper plate is a P $^+$ polysilicon plate with a platinum silicide serpentine across the top. During the test, current is passed through the silicided serpentine increasing the temperature of the upper plate. The figure shows the characteristic cross section of the structure.

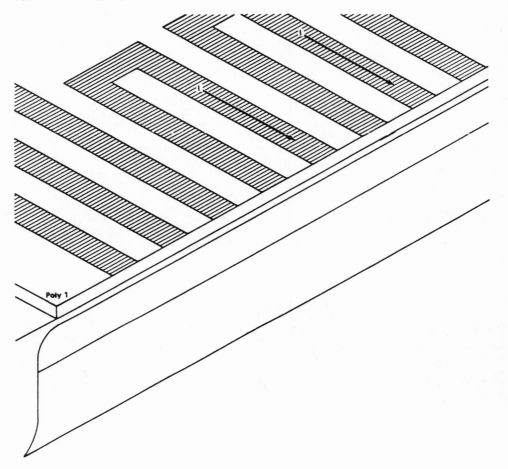

Cross Section of Mobile Ion Test Structure.

Since it has been shown that the ion mobility increases as temperature increases, the self–heated structure accelerates the associated mechanisms. The temperature and thus the required current is determined in the following manner. First, a linear relationship has been shown for resistivity vs. temperature for platinum silicide (PtSi), and a linear coefficient has been extracted. Using an external heat source temperature vs. resistance is determined. Second, using the resistant coefficient temperature vs. current can be accurate.

Thus, a minimum baseline is determined and C–V shifts of production material compared to it. Again, a simple pass–fail criterion is determined.

5. SCHOTTKY DIODE STRUCTURES

Schottky diodes [7] are critical devices in VLSI, not only as circuit elements but also as process characterization structures. Schottky diodes are useful in measuring the density of dopants (N_D) trap centers (N_T) and minority carrier lifetime. Although, other tests are available for limited use, such as spreading resistance or DLTS, they are either destructive or require complex sample preparation and sophisticated instrumentation. Wafer level measurement allows rapid nondestructive analysis on the product wafer.

A Schottky diode test structure, using self–heating for temperature measurement, is shown below. The Schottky structure is formed by opening a serpentine structure to epi and then forming a PtSi layer in the exposed silicon area. The diode is heated by forcing a current laterally through the PtSi region. During heating the devices can be biased for testing at temperature.

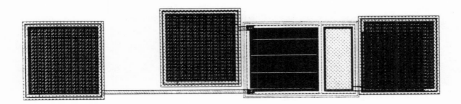

The Schottky Diode Test Structure.
Note the Serpentine PtSi Area.

Schottky diodes are very useful for measuring impurity concentration in doped silicon. When a Schottky diode is reversed biased, its depletion region spreads into the underlying lightly doped epi region. The capacitance of the reversed biased Schottky diode is a function of impurity concentration. The doping concentration can be determined from the relation in the following figure.

$$N_D = [2/(E_R E_0 q)][d/dv(1/C_{SC}{}^2)]^{-1} \qquad\qquad [6]$$

Where: $C_{SC}{}^2$ is the Schottky capacitance
q is charge
E_R is the dielectric constant
E_0 is the permitivity of free space

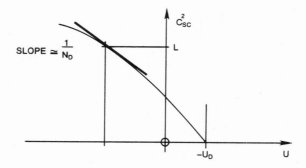

NoteThat the Doping Concentration Can Be Obtained from the C_{SC}^2 vs. U Curve.

If the Schottky diode is kept at a constant reverse bias and heated, the capacitance will change. If the experiment is performed at several heating rates, mid–band gap impurity characteristics can be measured.

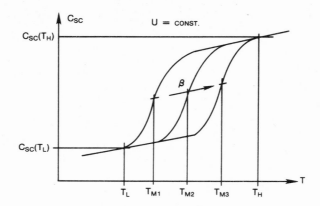

C_{SC} vs. Temperature Can Be Used to Obtain Mid Band Gap Impurity Characteristics.

$$N_T = [2(-V)/(qE_R E_0)][V_D(T_L) - V_D(T_H)][C_{SC}^2(T_H) - C_{SC}^2(T_L)] \qquad [7]$$

V_D is the voltage drop across the depletion region. From this an activation energy (E_T) can be determined:

$$E_T/KT_M = 1N[E_T/KB(1t2Kt_M/E_T)] \qquad [8]$$

where B is the rate of increase in temperature.

An important concern is the integrity of the sidewall in a double polysilicon bipolar device. It can be seen that mobile ion contamination in the oxides over the space charge region can cause junction degradation. The figure below shows the sidewall region under normal conditions. However, if ions are present, a leakage path can be generated in the junction as shown in the second figure on the next page.

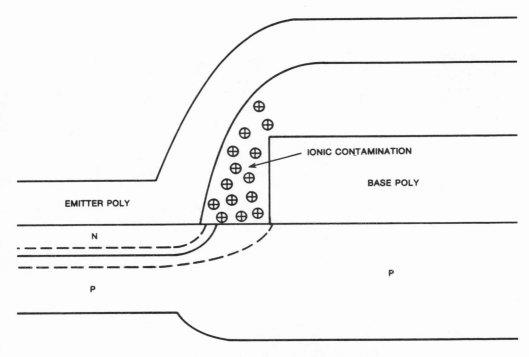

Ionic Contamination in an Unbiased Device.

6. WAFER LEVEL DEVICE RELIABILITY

Historically, bipolar device reliability has been of minimal concern; however, as scaling approaches submicron sizes and current densities increase, possible failure mechanisms must be investigated. The use of deposited oxides for isolation on the device level is a potential reliability concern.

As device spacings are reduced, ionic contamination can increase parasitic leakage current. Additionally, as device currents are reduced, these parasitic currents could significantly degrade device performance.

It is well known that most failure mechanisms are accelerated by temperature, voltage and current. Thus, a bipolar transistor with self–heating capability has been designed. The test structure, shown below, is a double polysilicon device. The emitter poly formed in an extremely long serpentine pattern with pad connections on each end. In this manner, current can be passed through the emitter, not through the device, directly heating the device junctions.

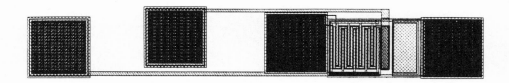

Bipolar Device Reliability Test Structure.
Note the Serpentine Emitter Can be Used to Heat the Device.

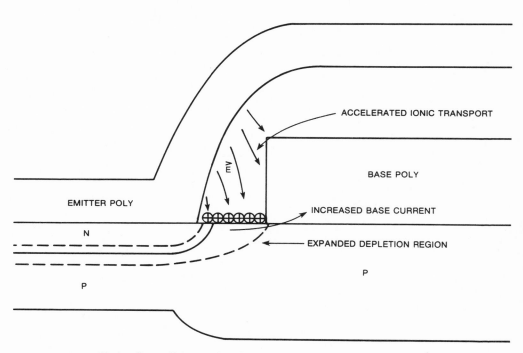

Biasing Causes Concentration of the Mobile Ions Creating a Leakage Path.

The structure shown in the figure* is also useful for basic device characterization. It is important to know device parameters over a wide temperature range. Typically wafers are heated on a hot chuck to make these measurements; with the heater structure the device can be heated to a known temperature and biased for characteristics, thus eliminating the need for an expensive hot chuck.

7. FUTURE WAFER LEVEL TESTING

It is clear that as the sophistication of semiconductor devices increases, wafer level testing must match that level. The future will show the level of integration of test structures increases. This increased sophistication will require drivers, sensors, and data collection to be implemented on the chip itself reducing the need for specialized off chip equipment. An added benefit to this is increased speed and ease of data collection. Thus, an entire series of tests could be multiplexed onto a small set of bond pads, increasing the usable scribe lane area.

An example of reliability integration is shown below where standard electromigration structures have been implemented.

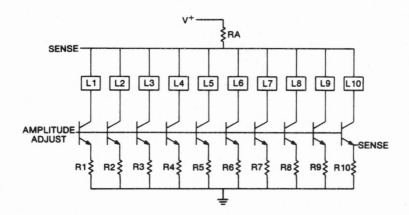

Wafer Level Test Structure of the Future.
Integration of Drivers, Sensing and Calibration Entirely Into the Device.

Ten lines can be stressed using five bond pads where 40 pads would have been necessary in earlier structures. Pad 1 is the supply voltage. Pad 2 senses voltage changes across R1 indicating individual line failures. Pad 3 is the current adjust input. Pad 4 is the ground. Pad 5 is for calibration.

Another circuit that evaluates contact electromigration implements a multiplexed output to sense the status of each contact (the contacts are in parallel and of varying sizes). High temperature AC measurement can also be implemented due to the stability and AC characteristics of advanced bipolar devices.

8. CONCLUSIONS

The pace at which reliability research advances must match the increasing sophistication of semiconductor devices. The transfer of emphasis from expensive and time–consuming package–level testing to wafer level testing will accelerate. The absolute necessity of this can be seen in the shorter product cycles, increased customization, and increasing performance vs. price characteristics. The future will see increased reliability integration, computer aided reliability and more accurate models of die level failure mechanisms.

* C_{SC} vs. Temperature Can Be Used to Obtain Mid Band Gap Impurity Characteristics.

REFERENCES

1   W. Meyer, D. Crook, (Intel), "A Non–Aging Screen to Prevent Wearout of Ultra–Thin Dielectrics", I.R.P.S. (1985).
2   B. Root, T. Turner, (Mostek), "Wafer Level Electromigration Tests for Productional Monitoring," I.R.P.S. (1985).
3   C. Hong, D. Crook, (Intel), "Breakdown Energy of Metal (BEM) – A New Technique for Monitoring Metallization Reliability at Wafer Level," I.R.P.S. (1985).
4   J. Black, (Motorola), Proceedings of the Third International Congress on Microelectronics, p. 141, Munich, (Nov. 1968).
5   E. Levine, J. Kitcher, (IBM), "Electromigration Induced Damage and Structure Change in Cr–Al/Cu and Al/Cu Interconnection Lines," I.R.P.S. (1984).
6   R. Thomas, D. Calabrese, (RADC), "Phenomenological Observations on Electromigration," I.R.P.S. (1983).
7   J. Zemel, Nondestructive Evaluation of Semiconductor Materials and Devices, Plenum Press, (1978).

WAFER LEVEL RELIABILITY TESTING: AN IDEA WHOSE TIME HAS COME

O.D. Trapp

Technology Associates
Portola Valley, California

Abstract. Wafer level reliability testing has been nurtured in the DARPA supported
workshops, held each autumn since 1982, at the Stanford Sierra Lodge on Fallen Leaf
Lake, Lake Tahoe, CA. The seeds planted in 1982 have produced an active crop of VLSI
manufacturers applying wafer level reliability test methods. Computer-Aided
Reliability (CAR) is a new seed being nurtured. Users are now being awakened by the
huge economic value of the wafer reliability testing technology.

Planting Seeds

In the late 1970's, NSA attempted to install wafer level reliability testing. The IC
manufacturers would not accept this concept. The idea of stressing test structures
to obtain their lognormal failure distribution was repugnant to say the least. Most
suppliers advised that they would not supply wafers if those kinds of tests were to
be done!

In the early 1980's, Paul Losleben moved from NSA to DARPA and again asked this
author to establish wafer level reliability testing for specific application for the
MOSIS program. The microelectronics industry has a long history of resisting ideas
forced upon them. Therefore, this author believed our industry should be nurtured in
the value of performing wafer level reliability testing.

The technical leaders of the IC industry were invited to send their key manufacturing
people to talk and think about these ideas in a workshop, open by invitation only.
Only U.S. companies were permitted to attend. Stanford University and University of
California (Berkeley) co-sponsored the workshops. These universities also contributed graduate students to work (for travel expenses) and participate in open, free
discussion with our industry technical leaders.

Initial Results

The first workshop concluded that although this was an interesting idea, it would not
work; it was just plain impractical; who would think of doing a probe test on the
wafer for 100's of hours, etc. But there was a glimmer of hope; there was a strong
agreement that the workshop should be held again.

The seeds did fall into fertile minds and ideas slowly became plans of actions. Why?
The time was right for this idea. With the increased drive for higher performance
VLSI devices, we were awakened to new reliability limitations. We were demanding
performance approaching the "Reliability Materials Limit" illustrated in Fig. 1. In
the 1960's and 1970's there was a wide "Margin of Reliability Assurance." Our designs and processing could be sloppy but the devices still yielded and were reliable.
But in the 1980's and beyond, the "Device Rules and Device Performance" will be
pushing up against these materials limits.

Our attention was focused of scaling algorithms essential for our moves from MSI to VLSI. The new failure mechanisms restricted and required modifications to these algorithms. Murray Woods of Intel frequently jolted our minds about the problems of 1 μm device reliability.

At the end of the second workshop, the question was not that wafer level testing could not be done, but where was it economic to do such tests? The seeds had germinated and the concept was healthy and growing.

New products use advanced design rules and new technologies. This is the ideal place to evaluate wafer level reliability testing. The results were staggering (Fig. 2). By applying reliability testing on the wafer, not on packaged devices, it is possible to rapidly solve reliability problems that are found to exist with the design rules and the processes. A normal qualification, as specified in MIL M 38510 or MIL STD 883, requires approximately 12 weeks to complete after the devices have been produced and assembled. By this technique, in the 1970's the average process/product development cycle time was 40 months. Today that has been shortened to an average of 30 months, for far more complex devices and processes.

By the end of the fourth workshop in 1985, it is clear that these ideas not only apply to process/product development, but are critically and economically important in high volume manufacturing.

There are still issues to be resolved. Can these wafer level reliability tests be correlated to traditional packaged reliability tests? Do both of these tests correlate to field reliability in VLSI devices? These questions must be addressed for each failure mechanism. The data reported at the 1985 workshop show that wafer level reliability tests do correlate to both packaged accelerated stress tests and to limited field data.

With or Without Wafer Level Reliability Testing

This comparison is complex (Fig. 3). There are many issues to consider. Each manufacturer and each user must understand the benefits and obligations of wafer level reliability testing. Figure 3 addresses the benefits. What are the obligations? Any new approach requires changes, new learning, acceptance of new values, job restructuring, etc.

If either device users or IC manufacturers value inexpensive, controlled manufacturing, then they will want to have wafer level reliability testing on their products (Fig. 4). For the U.S., this is KEY to our Strategic World Leadership. Today much is written about the fact that very few memory devices are made in the U.S.A. Should there be a national emergency that would separate us from our major sources of memory devices, we would be at a great disadvantage. By a broad implementation of wafer level reliability testing we can regain the necessary strategic role as VLSI leaders.

CAR

At the 1984 workshop we coined a new acronym, CAR. This stands for Computer-Aided Reliability. All industries have seen the benefits of CAD and CAM. Today designers can rapidly create very interesting, useful devices with the CAD tools. But in the area of reliability, few of the reliability engineers fully understand each of the failure mechanisms and their implications to the wide variety of VLSI designs. How then can we educate the multitude of designers so that they will not create monstrous reliability problems in future devices?

The only possible way to avoid future device chaos is to provide a CAR tool which can be integrated with CAD. This is not easy. Most of our failure mechanism mathematical models are crude at best. Most are tested by holding all variables constant except one. Unfortunately, devices have many parameters varying at the same time. CAR will cost money. But the value received will be even greater than the value of CAD. Will the U.S. accept the challenge or will it have to learn from other nations? Some activity is beginning, using internal funds, because our VLSI manufacturers know that the return justifies the investment. The workshop will continue to nurture this idea.

1986 Workshop

What is holding back the U.S. aerospace electronics industry? That is clearly a question for many to ponder. We encouraged aerospace users and manufacturers to participate actively in the 1986 Wafer Reliability Assessment Workshop. In the past only a few have attended. Does a meeting have to be visibly sponsored by a contracting agency to attract attention? Clearly that is important, but advances can also occur outside of funded meetings and funded activities!

The Wafer Reliability Workshops break the form of the traditional meetings. Clearly they have helped nurture a clear advance in reliability control and process control technology of doing accelerated life testing on the wafer. They are on the leading edge of the concept of CAR.

Margin of Reliability Assurance

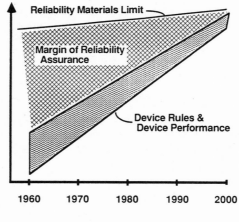

Figure 1

Process/Product Development Cycle Time

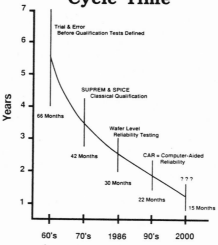

Figure 2

Wafer Level
Reliability Testing

<u>Without</u>	<u>With</u>

✦ Qual Test - 12 Weeks
      200 Devices
	✦ Qual Test - _On the wafer_
	      - hours/days

✦ Very Slow Corrective
    Action - Years
	✦ Fast Turn-Around
	    Corrective Action - Weeks

✦ Not Competitive Slow Development	✦ 30 months for 80386, etc.

✦ Delays &
    Cost Overruns
	✦ Key to "Just in Time"
	    Manufacturing

✦ Adverse Reliability
    Surprises
	✦ Controlled Device
	    Manufacturing - No Surprises

✦ Cost - Expensive	✦ Cost - Inexpensive
✦ Yields - Variable or Zero	✦ Yields - High

✦ Who Does It?
    MIL STD Users
	✦ Who Does It?
	    Commerical VLSI Makers &
	    Computer Industry

Figure 3

What is holding back the U.S. Aerospace Electronics Industry?

Figure 4

MICRO-FOCUS X-RAY IMAGING

Michael Juha

IRT Corporation
San Diego, California

INTRODUCTION

The acceptance of surface mounting in the electronics industry has been slowed by problems with component availability, electrical testing, and inspection. Industry suppliers and users have been working to solve these problems, and these problems are easing. Component manufacturers are offering substantially more devices in surface mountable packages. Automated test equipment vendors are offering test fixtures for surface mount circuit boards. And just recently, solder connection inspection has been automated through a "partnership" effort with a major electronics manufacturer. The result achieved in this successful first installation is the subject of this paper.

THE NEED FOR SOLDER QUALITY INSPECTION

Surface-mounted devices are held in place on the circuit board by their solder connections. The same situation occurs on plated-thru-hole circuit boards. However, with surface mounting, there is no solder plug surrounding a pin thru a hole in the circuit board to give the connection added strength. Instead, the solder alone bonds the device to the circuit board, as shown in figure 1. The electrical integrity of the circuit board is totally dependent upon the structural integrity of the solder connection. This issue of structural integrity, plus the large numbers of solder connections on each circuit board, is the impetus for automating solder connection inspection.

Surface mounting makes more electronic products man-portable or mobile in vehicles. With this portability or vehicular mobility comes shock, vibration, and extremes of temperature. Shock, vibration, and temperature approach or exceed levels that were previously associated only with military electronics. These higher stress levels must be sustained by the smaller solder connections characteristic of surface mounting. Numerous experts have cited how surface mounted solder connections are disposed to fatigue and creep failure.

STRESS AND ELECTRICAL TESTING

The structural integrity of solder connections can be assessed to some degree with "shake and bake" stress testing. However, this type of testing confronts the risk of wearing out the product before it ever reaches the customer. Further, the electrical testing that is used with "shake and bake", as well as that usually performed in regular manufacturing quality assurance, detects only the "open" or "short" conditions. Stress testing typically does not expose the structurally marginal connections that still conduct, but are long term candidates for failure, such as:

- o insufficient solder
- o poor wetting
- o excess solder or lead projection
- o device or lead off position
- o unwanted solder balls or splashes
- o device tilted relative to the board
- o porosity in the solder connection

VISUAL INSPECTION

Visual inspection can detect gross defects, such as missing devices, bridges outside the devices, the absence of solder fillets, and non-wetting. However, visual inspection is qualitative rather than quantitative---it does not measure the extent to which a defect exists. Also, visual inspection relies upon the external appearance of the solder connection to infer its internal structural integrity. And with surface mounting, the solder connections are partially or fully underneath the devices, making visual inspection impractical.

STRUCTURAL INSPECTION

Structural inspection is not a new problem. For years, aerospace and casting manufacturers have used X-ray inspection to examine the structural integrity of castings for airframes, engines, and transmissions. Since solder connections are a "casting" formed by the surface tension of the molten solder and the surfaces of the device and the circuit board, X-ray techniques will work for solder connections. The keys to making X-ray techniques viable for solder connection inspection are to:

1) speed up the X-ray imaging process
2) radiation from damaging electronic components
3) improve X-ray imaging to resolve 0.001 inch features
3) automate inspection to achieve fast, accurate results
4) make the techniques usable in the production line

DEVELOPING A SOLDER QUALITY INSPECTION MACHINE

In 1984, we became aware of the need for structural inspection of solder connections in surface-mounted electronics thru a customer, and developed the machine shown in figure 2 specifically for solder connection inspection. This machine took "flash" X-ray images of each device on the circuit board to keep radiation from damaging the device, and automatically inspected the "structure" of each solder connection according to a "rule set" that took into consideration:
o   the type of device (PLCC, SOT, LCC, etc.)
o   the shape of the pad on the circuit board
o   the amount of misalignment allowed between the
     device and the circuit board
o   the range of solder connection thickness allowed
o   the range of solder connection shape allowed
o   the amount of porosity allowed

This first machine administered a maximum dose of 5 RAD(Si) to each circuit board, was designed to "see" features as small as 0.002 inch in size, and used images like the one shown in figure 3 to inspect each solder connection. A schematic for this first machine is shown in figure 4. In operation, the machine used an electrical X-ray source to project a collimated beam of X-rays up through a 1" by 1" area of the circuit board. The X-ray shadow image of the solder connections was projected onto a fluorescent screen just above the circuit board. This screen converted the X-ray image into a visible light image, which was viewed by a high-resolution video camera through a first surface mirror to keep the camera and optics out of the X-ray beam. The video image of the solder connections was input to a digital image processor that performed the actual inspection under the direction of a set of programs in an IBM Personal Computer.

RESULTS ACHIEVED WITH THE FIRST MACHINE

In mid-1985, this first machine was installed in a surface mount production line to inspect 1000 circuit boards per day with J-leaded surface mounted ICs. Each circuit board had 252 solder connections, and inspection time per circuit board was 30 seconds. Since the customer's production line operated seven days per week, this first machine and its software have inspected more than 200,000 circuit boards $\approx 63 \times 10^6$ joints. This large production volume has forced us to make our inspection programs effective for the wide variations found in production solder connections made with vapor phase reflow.

The defects identified by this first machine have been (and are today):

o void (absence of solder joining lead to pad)
o insufficient solder (including porosity)
o bent lead (off position from pad)
o leads touching (producing a short without a bridge)
o solder bridge
o device off position (skewed or shifted)
o device missing from board

These defects are identified accurately and repeatably, and our customer is pleased with the performance of the machine (particularly since it has already paid for itself). However, there are some caveats on inspection accuracies.

Solder bridges and missing devices are practically always found, since they represent extreme conditions. Insufficients, bent leads, voids, and off positions are questions of the degree to which the defect is present. Through manual re-screening of automatically inspected circuit boards, we have learned that these defects are found roughly 95% to 99% of the time. The range from 95% to 99% is largely attributable to the variability of the human inspectors used to perform the re-screening. Inspectors do make "bad" calls on occasion. A more important facet is the relationship we found between increased defect detection and increased "false rejects", product rejected as bad when it is truly good.

THE IMPORTANCE OF ACCEPT/REJECT THRESHOLDS

Figure 5 shows two overlapping distributions that help explain this relationship. The horizontal axis is a "measure of quality" that is a composite of many measurements of the size and thickness of each solder connection. The vertical axis is the number of solder connections with that measure of quality in a batch of boards. The accept/reject threshold determines whether a solder connection is accepted as good or rejected as bad. Solder connections to the left of the threshhold are rejected as bad. Solder connections to the right of the threshhold are accepted as good.

This relationship between defect detection and false rejects can be seen in Figure 5. As the accept/reject threshold is moved to the right, more and more defects are detected until practically none escapes inspection. However, as defect detection grows in effectiveness, so does the number of good connections (the left "tail" of the "good" distribution) that will be falsely rejected as defective. This results from the overlap of the "good" and "bad" solder connection distributions, and reflects reality. We have found that the characteristics of marginally good solder connections significantly overlap those of marginally bad solder connections.

As a result of the relationship between the accept/reject thresholds and the economics of our customer, the accept/reject thresholds for the first machine have been set up to detect roughly 97% of all defects while making fewer than 5% false rejects. These performance levels are far better than those achievable with inspection personnel. And, inspection by our machine is done before electrical testing to increase the effectiveness of electrical tests. For other customers with a different manufacturing process, and different costs for inspection, rework, scrap, and escape of defects, different accept/reject threshholds would be necessary to achieve the best economic return for their circumstances.

THE "STRUCTURAL" SOLDER QUALITY STANDARDS PROBLEM

As an aside comment, the requirement for flexible accept/reject thresholds, when coupled with the wide variations found in production solder connections, and the absence of complete information about what a good solder connection "looked like", almost prevented us from delivering a satisfactory working machine. Fortunately, our customer was willing to spend considerable time and money developing their own structural standards for what made solder connections good versus bad. This required stress cycling hundreds of circuit boards, analyzing each failure to establish causes, and then proceeding with production while monitoring production items for in-the-field failures on an on-going basis. All this work entailed considerable investment, and resulted in standards that are not consistent with present visual inspection standards. When structural standards are developed for other products, such as avionics, we feel these standards will not agree with existing visual inspection standards. Since a substantial beneficiary of these new structural standards would be the military, funding for standards development should be allocated as soon as possible, particularly in view of the increasing concerns over the structural viability of leadless surface mounted devices. Based on the experience of our customer, where practically no field failures now occur, these new standards would clearly help reduce in-service failures. And, our customer's product environment is 3+ G's of shock and vibration, ambient temperature from -40 F to +125 F, and humidity from 0% to 100%.

CONCLUSIONS

It is difficult to extrapolate general savings rules from a single installation. However, we have shown with our first installation that automated X-ray inspection can dramatically reduce:

o the costs of inspection
o the incidence of unnecessary rework on good boards
o the recycling of boards thru rework as additional
   defects are cited
o the costs of scrap by minimizing rework
o the escape of defective boards
o the incidence of defects

This last point is an often overlooked major benefit area. With the quantitative quality data that is a by-product of automated X-ray inspection, you can control your manufacturing process to make a better product. Our first machine was installed in a new manufacturing line with completely new equipment. During process start-up, it was discovered that our machine could help set up the solder paste screen printer and reduce the incidence of voids, bridges and insufficients. During production, our machine continues to monitor paste printer performance by noting the incidence of bridges and insufficients. When our defect reports show an increase in bridges or insufficients, the customer's personnel know how to adjust the process back into control. As a result, our customer has been able to achieve a significant increase in yield.

Figure 1.   Visual image of PLCC on a circuit board.

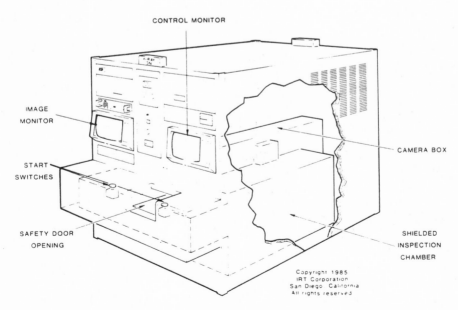

Figure 2.   Line drawing of first inspection machine.

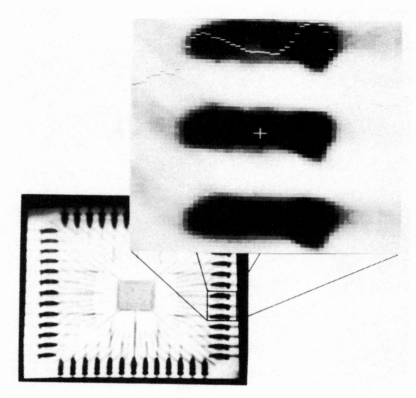

Figure 3.   X-ray image of PLCC.

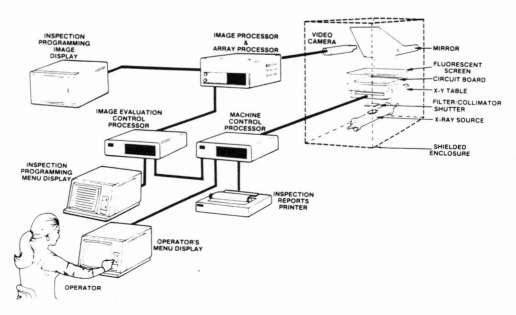

Figure 4.   Schematic for first inspection machine.

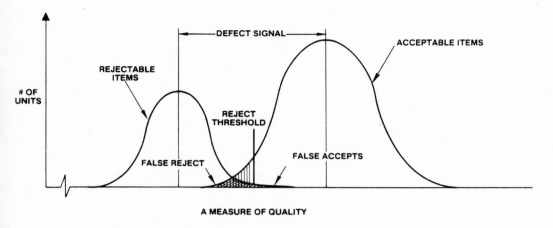

Figure 5.   Distributions showing solder connection quality.

MEASUREMENT OF OPAQUE FILM THICKNESS

R L. Thomas, J. Jaarin,* C. Reyes, I.C. Oppenheim, L.D. Favro, and P.K. Kuo

Department of Physics
Wayne State University
Detroit, Michigan

INTRODUCTION

We describe the theoretical and experimental framework[1-6] for thickness measurements of thin metal films by low-frequency thermal waves. Although we assume that the films are opaque and the substrates are comparatively poor thermal conductors, the theory is easily extended to other cases of technological interest. We begin with a brief description of thermal waves and the experimental arrangement and parameters. Next we illustrate the usefulness of the technique for making absolute measurements (based on measurements of length and time) of the thermal diffusivities of isotropic substrate materials. This measurement on pure elemental solids provides a check on our three-dimensional theory in the limiting case of zero film thickness. The theoretical framework is then presented, along with numerical calculations and corresponding experimental results for the case of copper films on a glass substrate.

DESCRIPTION OF THERMAL WAVES AND EXPERIMENTAL TECHNIQUE

The elements of thermal wave propagation are illustrated by considering the one-dimensional heat equation, presented along with its solution in Fig. 1. Here k is the thermal conductivity, ρ is the mass density, c the specific heat capacity, and ω is the angular frequency of the (assumed) periodic heat source. By inspection, one sees that the solution is a wave whose wave vector is complex, and has equal real and imaginary parts. A sketch of the spatial variation of the real part (see Fig. 1) emphasizes an important aspect of thermal wave propagation - these waves are nearly completely damped out after propagating one thermal wavelength. Both the thermal wavelength and the proportional quantity, μ_s, the thermal diffusion length of the solid, are seen to be inversely proportional to the square root of the heat source frequency. At the typical ranges of thermal properties and experimental frequencies, diffusion lengths range from a few micrometers to a few millimeters.

The arrangement for the thermal wave measurements described here is shown in Fig. 2. The heating beam is an intensity-modulated Ar+ ion laser, focussed to a few micrometer diameter spot on the sample surface. For the present measurements, frequencies are typically below 1 kHz, such that the thermal diffusion lengths in the Cu films are much greater than their thicknesses. The time-varying temperature distribution in the air just

---

*Dept. of Physics, University of Helsinki, Helsinki, Finland.

above the coating is monitored phase-coherently by means of a vector
lock-in amplifier, which measures the time-varying deflection of a HeNe
probe laser beam skimming the surface of the coating (see Figs. 2-4,
illustrating the use of the mirage effect in the heated air). In this
experiment, the probe beam is fixed in position and the heating beam spot
is scanned (transverse offset) across the surface beneath the probe beam
and at right angles to it. Figure 5 shows the resulting in-phase
component of the transverse probe beam deflection (i.e. the deflection
component in a plane parallel to that of the sample surface) during such a
scan. Note that the transverse deflection is zero as the heated spot
passes directly beneath the probe beam, and that its sign changes at that
point. The length measurement for determining both the film thickness and
(accompanied by the frequency measurement) the thermal diffusivity of the
substrate, is the quantity x_0 in Fig. 5, namely, the separation between
the two non-central zero crossings of the in-phase signal.

THERMAL DIFFUSIVITY OF AN ISOTROPIC SOLID

A plot of x_0 versus the reciprocal of the square root of the
frequency should yield the thermal diffusivity, $\alpha = k/\rho c$. Experimental
verification of this fact for pure elemental solids is given in Figs. 6
and 7. Here, the nominal diffusivity is determined from the Handbook of
Chemistry and Physics values for k, ρ and c.

The length measurements used in the preceding figures employed only a
few data points from the scan. As a further check on the reliability of
the theory, we plot the in-phase component of the transverse deflection
versus the quadrature component of that deflection, during the scan of
transverse offset (see Fig. 8). The resulting comparison between theory
and experiment for the case of silver (see Fig. 9) uses no adjustable
parameters (the Handbook value for diffusivity is assumed), and shows
excellent agreement.

THEORETICAL FRAMEWORK FOR THIN FILM CALCULATIONS

The geometry for the thin film experiment is given in Fig. 10. The
thickness of the film is a, the radius of the heating beam is b, that of
the probe beam is c, the height of the probe beam is h_0, the transverse
offset is y_0, and κ, with its appropriate subscript, is the thermal
conductivity. The thermal conductivity of the air is assumed to be
negligible compared to those of the film or substrate. The theoretical
equation is given in Fig. 11. A detailed description of this theory is
found elsewhere.[1]

NUMERICAL CALCULATIONS AND EXPERIMENTAL RESULTS: Cu FILM ON GLASS

Numerical calculations of x_0 as a function of inverse root frequency
for Cu films (1000 A to 5000 A) on glass are shown in Fig. 12, and the
corresponding experimental measurements are shown in Fig. 13. Theory and
experiment are in excellent agreement. Figure 14 shows the theoretically
predicted dependence of x_0 on coating thickness for three different
frequencies. The film thicknesses were also measured independently by

means of Rutherford backscattering of alpha particles. The two measurements agree to within a combined uncertainty of about 10%.

SUMMARY AND CONCLUSIONS

We have described a thermal wave technique which is capable of determining the thicknesses of opaque metal films on substrates whose thermal diffusivities are small compared to those of the films. The method is based on measuring the transverse deflection of an optical probe beam, due to the mirage effect in the air above the sample, as a function of the transverse probe beam distance from a localized ac surface heat source. The measurement is carried out in the frequency range below 1 kHz. by fitting the data to the theory of Kuo et al.[1], without prior knowledge of the diffusivities or the conductivities of the coating or the substrate, one can determine the thickness of the film as well as the thermal diffusivity of the substrate. We have applied this method to copper films on glass. The thicknesses of these films were between 1000 A and 5000 A. We find agreement between the thicknesses determined by our method and by measurements of the Rutherford backscattering of alpha partices, carried out in this laboratory, to within a combined uncertainty of approximately 10%.

ACKNOWLEDGEMENTS

This work was supported in part by ARO under Contract No. DAAG 29-84-K-0173, and in part by the Center for Advanced Nondestructive Evaluation, operated by the Ames Laboratory, USDOE, for the Air Force Wright Aeronautical Laboratories/Materials Laboratory under Contract No. W-7405-ENG-82 with Iowa State University. J. Jaarinen also acknowledges the support of Finland's Cultural Foundation.

REFERENCES

1. P.K. Kuo, E.D. Sendler, L.D. Favro and R.L. Thomas , Can. J. Phys., Vol. 64, No. 9, (September 1986), 1168.

2. P.K. Kuo, C.B. Reyes, L.D. Favro, R.L. Thomas, D.S. Kim, and Shu-Yi Zhang, Review of Progress in Quantitative NDE, Vol. 5B, edited by D.O. Thompson and D. Chimenti (Plenum, New York, 1986), 1519.

3. R.L. Thomas, L.D. Favro, D.S. Kim, P.K. Kuo, C.B. Reyes, and Shu-Yi Zhang, Review of Progress in Quantitative NDE, Vol. 5B, edited by D.O. Thompson and D. Chimenti (Plenum, New York, 1986), 1379.

4. R.L. Thomas, L.J. Inglehart, M.J. Lin, L.D. Favro, and P.K. Kuo, Review of Progress in Quantitative NDE, Vol. 4B, edited by D.O. Thompson and D. Chimenti (Plenum, New York, 1985), 859.

5. P.K. Kuo, M.J. Lin, C.B. Reyes, L.D. Favro, R.L. Thomas, D.S. Kim, Shu-yi Zhang, L.J. Inglehart, D. Fournier, A.C. Boccara, and N. Yacoubi, Can. J. Phys., Vol. 64, No. 9, (September 1986), 1165.

6. P.K. Kuo, L.J. Inglehart, E.D. Sendler, M.J. Lin, L.D. Favro, and R.L. Thomas, Review of Progress in Quantitative NDE, Vol. 4B, edited by D.O. Thompson and D. Chimenti (Plenum, New York, 1985), 745.

THERMAL DIFFUSION WITH A PERIODIC SOURCE
(ONE-DIMENSIONAL PICTURE)

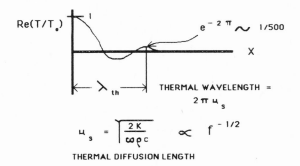

DIFFUSION EQUATION

$$k \frac{d^2 T}{dx^2} = \rho c \frac{\partial T}{\partial t}$$

SOLUTION:

$T = T_\bullet \exp[i(qx - \omega t)],$ where

$$q = \frac{1+i}{\sqrt{2}} \sqrt{\frac{\omega \rho c}{k}}$$

$Re(T/T_\bullet)$

$e^{-2\pi} \sim 1/500$

X

THERMAL WAVELENGTH = $2\pi \mu_s$

$$\mu_s = \sqrt{\frac{2K}{\omega \rho c}} \propto f^{-1/2}$$

THERMAL DIFFUSION LENGTH

Figure 1

ARRANGEMENT FOR THERMAL
WAVE SCAN OF A COATING

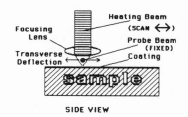

SIDE VIEW

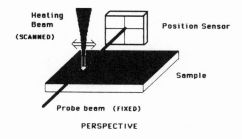

PERSPECTIVE

Figure 2

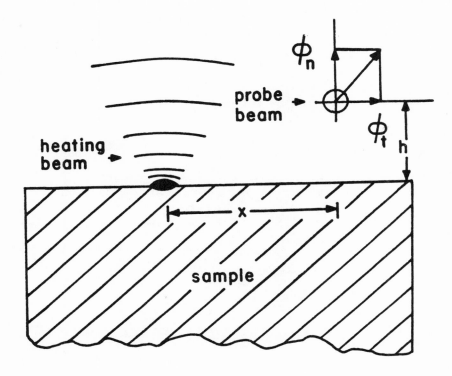

Figure 3

Mirage Effect (Optical Beam Deflection) Detection

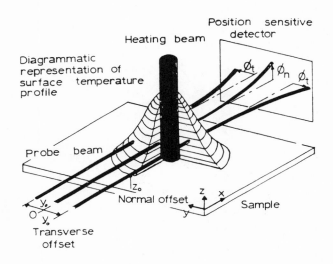

After L.C. Aamodt and J.C. Murony, J.Appl Phys. 54 581 (1983).

Figure 4

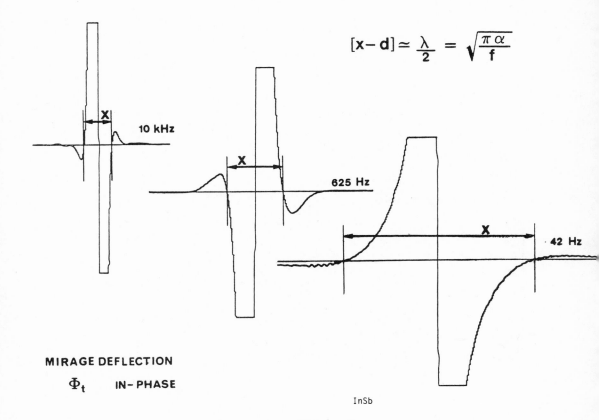

$$[x-d] \simeq \frac{\lambda}{2} = \sqrt{\frac{\pi\,\alpha}{f}}$$

10 kHz

625 Hz

42 Hz

MIRAGE DEFLECTION

Φ_t IN-PHASE

InSb

Figure 5

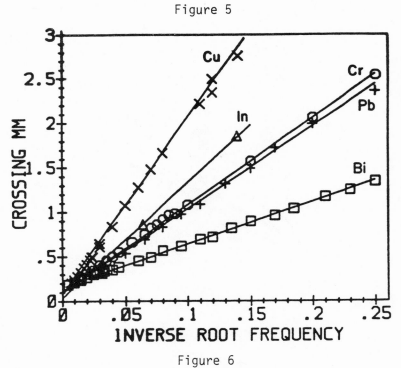

Figure 6

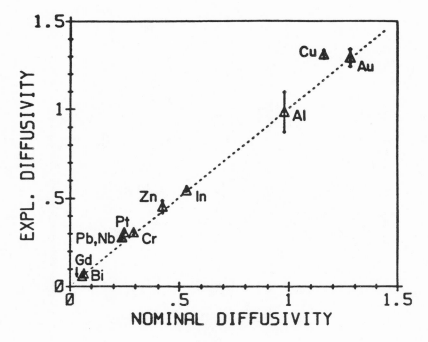

Figure 7

TOP VIEW

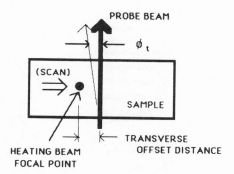

AS THE HEATING BEAM FOCAL POINT
IS SCANNED ACROSS THE SAMPLE:

PLOT

IN-PHASE COMPONENT OF ϕ_t

VERSUS

QUADRATURE COMPONENT OF ϕ_t

Figure 8

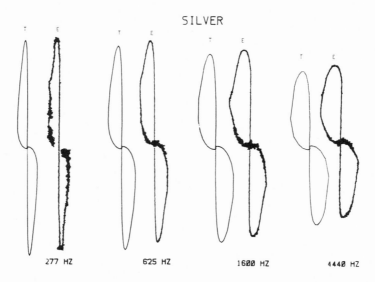

Figure 9

GEOMETRY

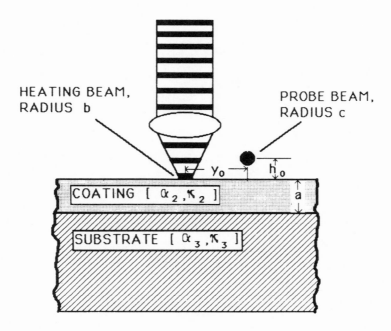

Figure 10

THEORY FOR THE MIRAGE TRANSVERSE
DEFLECTION SIGNAL (S)

$$S(y_o, h_o) = \frac{1}{c^2} \left[\frac{-i\omega c^2}{4\alpha_1} \right] \cdot$$

$$\cdot \int_0^\infty \frac{k\,dk}{\sqrt{k^2 - \frac{i\omega}{\alpha_2}}} \sin ky_o \exp\left\{ -\frac{k^2 b^2}{4} - h_o \sqrt{k^2 - \frac{i\omega}{\alpha_1}} \right\} \cdot$$

- $F(k)$, where

  b = Gaussian beam radius of heating beam

  c = Gaussian beam radius of probe beam

  α_1 = Thermal diffusivity of gas (region 1)

  α_2 = Thermal diffusivity of coating (region 2)

  α_3 = Thermal diffusivity of substrate (region 3)

$$F(k) = \frac{1+R}{1-R} \ , \ R = \frac{1-R'}{1+R'} \ \exp\left[-2a\sqrt{k^2 - \frac{i\omega}{\alpha_3}} \right]$$

$$R' = \frac{\mathcal{K}_3 \sqrt{k^2 - \frac{i\omega}{\alpha_3}}}{\mathcal{K}_2 \sqrt{k^2 - \frac{i\omega}{\alpha_2}}} \quad ; \ \mathcal{K} = \text{Thermal Conductivity,}$$

$$a = \text{Film Thickness}$$

Figure 11

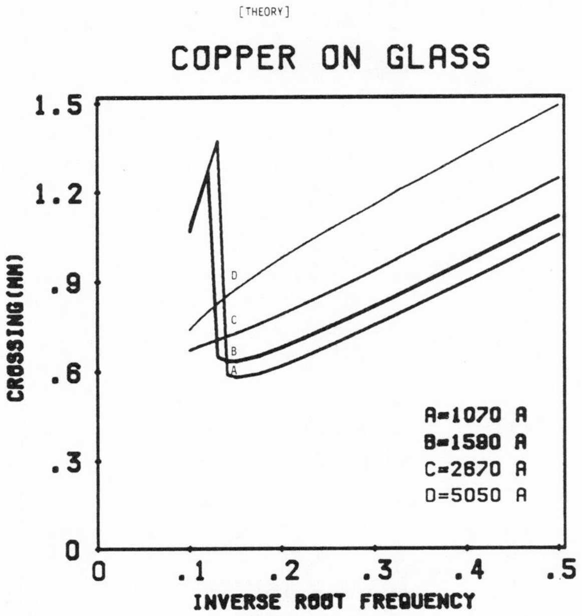

[THEORY]

COPPER ON GLASS

A=1070 A
B=1590 A
C=2870 A
D=5050 A

Figure 12

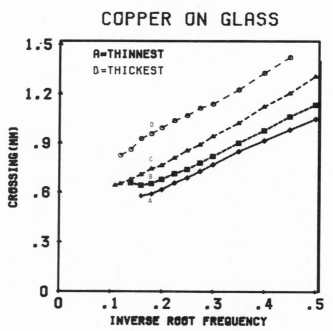

Figure 13

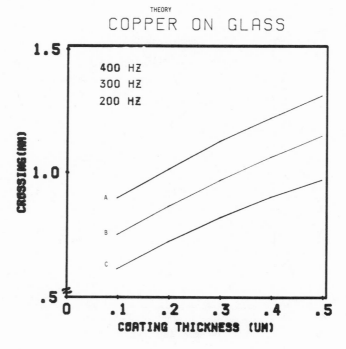

Figure 14

INTELLIGENT LASER SOLDERING INSPECTION AND PROCESS CONTROL

Riccardo Vanzetti

Vanzetti Systems, Inc.
Stoughton, Massachusetts

INTRODUCTION

Component assembly on printed circuitry keeps making giant strides toward denser packaging and smaller dimensions. From single layer to multilayer, from through-holes to surface-mounted components (SMDs) and tape-applied bonds (TAB), unrelenting progress results in new, difficult problems in assembling, soldering, inspecting and controlling the manufacturing process of the new electronics.

Among the major problems are the variables introduced by human operators. The small dimensions and the tight assembly tolerances are now successfully met by machines which are much faster and precise than the human hand. The same is true for soldering. But visual inspection of the solder joints is now so severely limited by the ever-shrinking area accessible to the human eye that the inspector's diagnosis cannot be trusted any longer. It is a slow, costly, unreliable and often misleading relic of the pre-automation era. As a matter of fact, it is the only operation still being performed by humans. A solution must be found to fill this gap.

NOVEL APPROACH: THERMAL FLOW

The solution to the problem of assessing the quality of a soldered joint is based on monitoring how heat flows through the joint itself. Evidently, heat injected at one end of the joint will spread through it, reaching the other end faster or slower, according to the quality of the heat transfer path. Any obstacle along its path (voids, inclusions, discontinuities, etc.) will slow down the heat transfer from one end to the other. Figure 1 shows how this approach is applied to a "lap-joint."

A measured pulse of laser radiation is injected on the surface of a "gull-wing" wire soldered to a pad. An infrared detector measuring the temperature at the point of heat injection will "see" a temperature rise during the laser heating pulse and a temperature decay after it. The analog signal at the detector's output is called the "infrared signature" or "thermal signature" of the corresponding solder joint. It contains all the information needed to define the quality of the joint. This is because the shape of the signature is affected by the following four variables affecting the joint:

a) surface cleanliness

b) surface emissivity

c) thermal mass

79

d) heat sinking

Item a) is mainly reflected in the initial "rise" of the signature, and it can be caused by residual flux or any kind of deposited material or film on the joint's surface. It often results in mini-fires, so that sometimes the laser beam is automatically turned off prematurely.

Item b) gives information about cold solder joints, tin depletion of the solder bath, presence of contaminants (gold, copper, iron, etc.) in the solder alloy, and excessive intermetallic formation.

Item c) indicates either excess of solder material or insufficiency of it (because of voids, dewetting and the like).

Item d) (located at the tail end of the signature) indicates whether the solder joint is properly connected to the heat-sinks, such as the component's lead and the printed wiring leading away from the pad.

Figure 2 shows typical signatures related to some of the conditions listed above.

THE LASER/INSPECT SYSTEM

The schematic diagram of the system which developed these signatures is shown in Figure 3. It is called "Laser/INSPECT" and it utilizes a 30 watt YAG laser as the heat source, together with a 0.5 milliwatt Helium-Neon laser (coaxial with the YAG) which illuminates with visible light the point of the target being heat injected. The infrared detector is an In-Sb photovoltaic cell cooled at 77° Kelvin and made blind to the lasers' wavelengths, so that it will only receive the blackbody (or graybody) radiation between 2.5μm and 5.5μm emitted as a function of temperature by exactly the same area heated by the laser.

The p.c.b. under inspection is mounted on a very fast and precise X-Y table controlled by the computer, which is programmed to bring in rapid succession each solder joint under the focal point of the optical head. The computer also opens and closes the laser shutter and processes the signature information arriving from the detector after analog-to-digital (A/D) conversion. The computer's memory holds, for each joint, the standard signature which is used as reference against which to compare the signature of the joint being inspected. The results are printed out by a fast printer, so that every joint has its own inspection certificate on a hard copy printout.

A TV camera extracts the visible image of the area being inspected, and can be used for programming the X and Y coordinates of the joints to be inspected and also for watching the inspection process during operation.

Figure 4 shows the Laser/INSPECT system and its crew. From left to right we see the programmer, standing in front of the closed inspection compartment, containing the lasers, the optical head, and the X-Y table with the p.c.b. under inspection. The laser power supply console and the computer console are under the inspection compartment. In the center, the systems operator sits in front of the monitor with its keyboard. At far left the printouts are being read to look for indications of defective solder joints.

In Figure 5 we see how the laser beam is focused on a solder joint to be inspected. According to the type of joint to be inspected, there is a "best angle" at which the laser beam can strike it most efficiently. Accordingly, the optical head can be tilted in the four directions +X, −X, +Y, and −Y.

And what about the speed of operation? For plated-through holes, the laser pulse can take between 60 and 100 milliseconds. To these we'll have to add another 15 milliseconds for taking three readings during the signature decay, plus 50 milliseconds for the table movement to bring the next joint at the focal point of the optical system. The total time adds up to 125 or 165 milliseconds per joint, or between 8 and 6 joints/second, for plated-through holes. For the much smaller SMD joints, a laser pulse of 25 milliseconds is adequate, so that the total time to inspect one joint and move to the next position can be reduced to 90 milliseconds, which results in an operational speed of 11 joints/second. This is certainly faster than any visual inspection deserving such an appellative, since this could vary between 5 joints/second and 5 joints/minute, according to the quality requirements of the electronics to be inspected.

STATISTICAL DATA

The information supplied by the Laser/INSPECT (L/I) system can be elaborated and outputted by the computer in several different formats. A comparative evaluation program carried out by Texas Instruments is worth mentioning. In order to obtain permission by the U.S. Navy to use the Laser/INSPECT system instead of visual inspection for the electronics p.c. boards of the HARM (High-Velocity Antiradiation Missile) production, Texas Instruments (TI) ran 186,570 solder joints first through the L/I system and then through conventional visual inspection (accordingly to the prescribed MIL-SPECS).

Figure 6 shows the result of the final comparison of the two different inspection approaches. For only 57% of the joints there is agreement between visual and the Laser/INSPECT. For the remaining 43% there is total disagreement (25% visual accepts, L/I rejects; 18% visual rejects, L/I accepts). Subsequent microsectioning of the "disagreed" joints proved in every instance that the L/I diagnosis was correct, as opposed to the visual diagnosis.

On the basis of this evidence, the U.S. Navy authorized TI to use the L/I system instead of the visual inspection mandated by the MIL-SPECS. As a consequence, today TI delivers to the Navy better quality HARM electronics, while the cost of each missile has been cut in half.

Histograms are useful in processing the L/I date. They offer a quick way to verify whether the soldering process is within tight control or drifting out of it. Figure 7 is an impressive example of the large range of variations introduced by the human hand in the soldering operation.

Figure 8 shows more examples of such histograms. In all these charts, the ordinate scale indicates in arbitrary numbers the value of the peak radiation of the infrared signature, while the abscissa indicates how many times each of those values was met by the signatures of the joints under test.

INSPECTING TAB ASSEMBLIES

The Laser/INSPECT technology is applicable to most types of joints or bonds. Both the sizes of the laser beam and of the detector spot can be adjusted to meet the target dimensions. Figure 9 shows a TAB assembly whose joints were inspected by a microscopic version of the Laser/INSPECT, in which the Laser spot is .022" (or 0.05mm) and the detector focal area .044" (or 0.10mm). Figure 10 shows some of these joints.

It can be seen that seven of the joints were prevented from making electrical contact by some epoxy smear. This lack of contact is clearly reflected in the high peaks of their signatures, which contrast with the low peaks of the adjacent wires, both at left and at right in the oscilloscope display picture of Figure 11.

INTELLIGENT LASER SOLDERER

From inspection to reflow soldering the step is quite short. Just a few more milliseconds of exposure to the laser beam, and the solid solder becomes liquid. During the change of phase, the temperature won't change. Figure 12 shows an oscilloscope display of such transition, where time runs along the abscissa and temperature rises along the ordinate. The two changes of phase, from solid to liquid (at left) and from liquid to solid (at right) are indicated by the two plateaux pointed by the arrows. These plateaux are inclined instead of horizontal, because in the area viewed by the detector there is simultaneously solid and liquid material.

Ad hoc software processes the detector output and turns off the laser shortly after full liquefaction is observed. In this way every joint receives the right amount of heat, through a "custom-tailored" laser pulse precisely measured to the individual joint's needs. This is because no two joints are identical. They differ in thermal mass and in heat sinking, so that their heat requirements to achieve liquefaction are different.

This is how the Laser Reflow Soldering System works, and this is why it is called intelligent.

Besides intelligence, the system has speed. On SMD's soldering can proceed at an average speed of 4 joints/second ("average", since every joint will need a different dwell time, as dictated by the individual heat requirement).

This 4 joints/second speed might appear slow when compared with the 1000 joints/minute typical of the mass-soldering systems today in use. But in fact it is much faster BECAUSE THE JOINTS ARE ALREADY INSPECTED. Their infrared signature is already their inspection certificate.

This means that the overall speed of the combined soldering and inspection operation is faster for the laser approach.

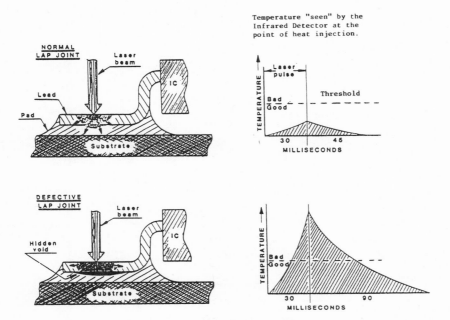

Figure 1. Behavior of normal and defective joints during laser/thermal testing.

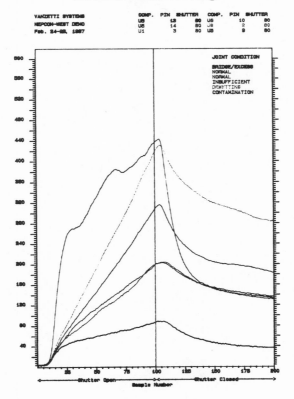

Figure 2. Typical infrared signatures of good and defective solder joints.

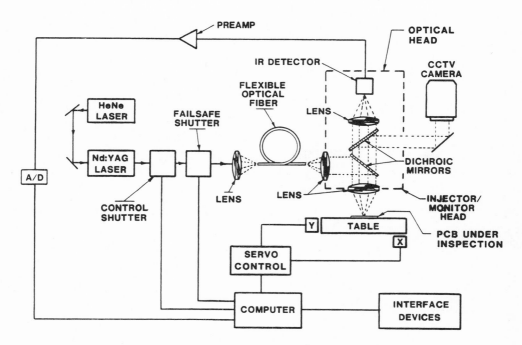

Figure 3.   Diagram of Laser/INSPECT system.

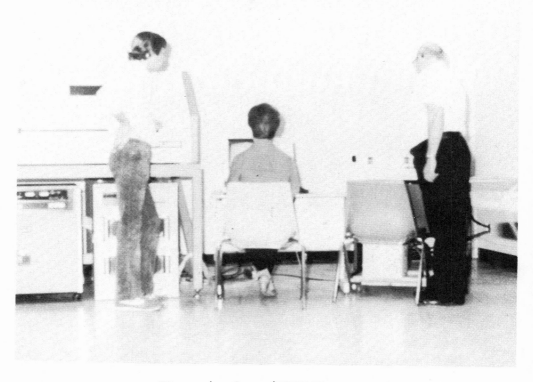

Figure 4.   Laser/INSPECT system.

Figure 5. Laser beam exiting from tilted optical head.

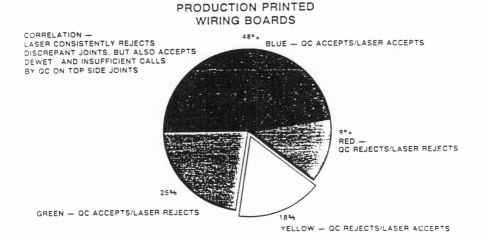

Figure 6. Comparing visual inspection versus Laser/INSPECT;
summary of 184,000 solder joints.

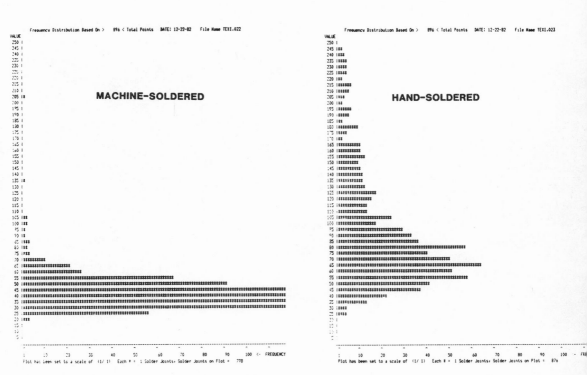

Figure 7.   Comparing soldering processes: machine versus manual.

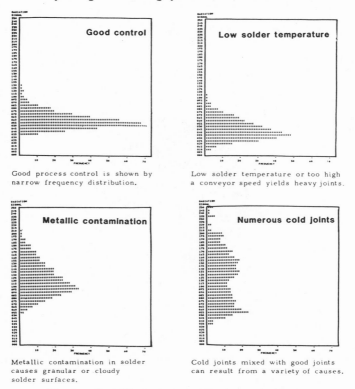

Good process control is shown by narrow frequency distribution.

Low solder temperature or too high a conveyor speed yields heavy joints.

Metallic contamination in solder causes granular or cloudy solder surfaces.

Cold joints mixed with good joints can result from a variety of causes.

Figure 8.   Histograms disclose drifting variables in soldering process.

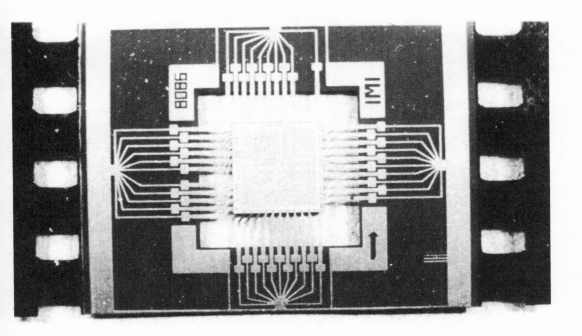

Figure 9.   TAB assembly.

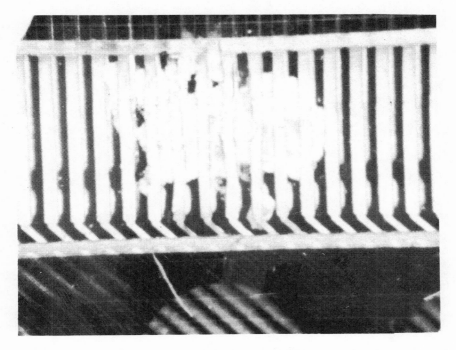

Figure 10.   Detail of figure 9, with seven unbonds.

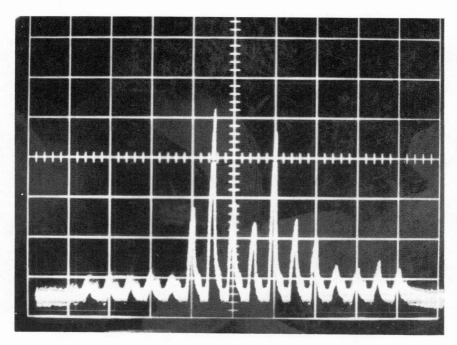

Figure 11.   Seven unbonds surrounded
by nine good bonds.

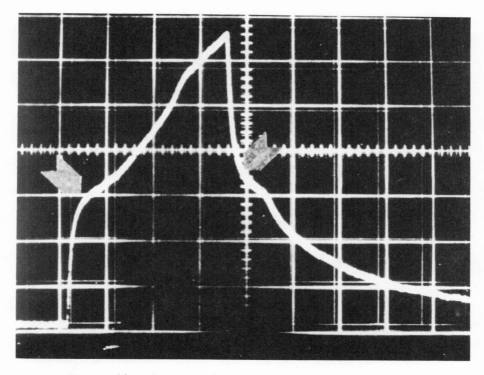

Figure 12.   Solder reflow: solid to liquid to solid.
Temperature diagram.

RUPTURE TESTING FOR THE QUALITY CONTROL OF ELECTRODEPOSITED COPPER INTERCONNECTIONS IN HIGH-SPEED, HIGH-DENSITY CIRCUITS

Louis Zakraysek

General Electric Company
Syracuse, New York

Introduction

The performance requirements of copper-clad laminates and printed wiring boards are driven by advances in integrated circuit technology and by the resultant miniaturization of components. The trends in this technology have resulted in sophisticated multilayer versions of the basic laminate structure. Some examples of high density printed wiring multilayer boards (PWMLBs) are shown in Figure 1.

Figure 1. High-density printed wiring multilayer boards.

The most recent impetus for performance improvement stems from the desirability for the surface mounting of components, from the electrical and thermal constraints imposed by high density component assembly, and by high-speed circuit design. These advances make it imperative that we construct PWMLBs with predictable and consistent properties in order to meet performance needs.

Recent experience[1,2] in using advanced, thermally stabilized multilayer substrates for high density assemblies shows that they are susceptible to premature failure due to the microcracking of the interconnecting conductive traces, including both the inner layer foil and the plated-through-hole (PTH). In some cases, fracture is initiated by a thermal stress, and the microcrack propagates during thermal cycling. In other cases, failure occurs by a tensile over-stress or by creep-rupture. For either failure mechanism, the result is an open circuit condition and the loss of a completed, and usually expensive, PWMLB.

Shown in Figure 2 are the three usual locations where microcracking will occur, namely, at the PTH corner, in the PTH barrel, and at an inner layer. Examples of these fractures taken from PTH microstructures are shown in Figure 3. Fractures such as those shown here are caused by an excessive tensile stress during thermal shock, by creep-rupture due to a sustained thermal stress or by fatigue during thermal cycling. The fracture surfaces are typically those of brittle materials, i.e., little or no evidence of plastic deformation.

Whether or not fracture actually occurs in a given PWMLB appears to be dependent on three factors: (1) a Z-direction thermal stress of sufficient magnitude to initiate fracture, (2) design- or process-induced stress risers, and (3) the presence of electrolytic copper with low hot strength. To date, most of the attention with regard to microcracking has been given the first two. In this paper we describe a modified rupture pressure testing procedure by which the mechanical properties of copper conductors in PWMLBs can be quantified with respect to strength and ductility at elevated temperatures. Since this information is obtained before the copper is used in a PWMLB, it can be used to prevent the use of a grade of electrolytic copper that is susceptible to premature thermal stress failure. In addition to eliminating failure under some conditions, this approach provides some latitude with regard to the handling of the other two factors that are mentioned as contributors to microcracking.

Type of Interconnecting Material

A printed wiring board (PWB) is the primary method used in the electronics industry for the interconnection of circuits. A large share of the PWBs in use are produced by the fabrication of copper-clad laminates from which the copper is selectively removed by a photolithographic process to define a circuit interconnection pattern. Most of the copper used as conductor traces is supplied to laminators by a relatively small number of producers who make it in foil form by electrolytic deposition. The laminator then applies this foil to a dielectric substrate for subsequent sale to the PWB fabricator.

When the user of the laminates completes the fabrication of a multilayer printed wiring board, the circuit traces on each of the individual layers are interconnected by a plated-through-hole (PTH) process. Therefore, all of the conductors on a structure such as this are made of electrolytic copper, the surface and inner layers being done by an outside source, and the PTH deposition being done by an in-house plating operation. In this paper, either of these conductors will be referred to as Type E copper. The industry standard for Type E copper foil is IPC-CF-150 (Institute for Interconnecting and Packaging Electronic Circuits), and in this standard are documented the various classes of foil according to their mechanical-physical properties. The classes that are of most interest for our purpose, are Type E, Class 1; Type E, Class 3; and Type W, Class 7. They are, respectively: electrolytic, ordinary quality; electrolytic, high-temperature ductile; and rolled, annealed.

It should be noted that rolled foil (Type W, Class 7 in the above-mentioned standard) is also used on PWB laminates. For a number of reasons, its use is limited mostly to special applications such as

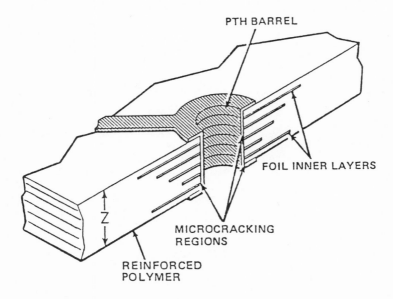

PTH BARREL

FOIL INNER LAYERS

MICROCRACKING
REGIONS

REINFORCED
POLYMER

Z

Figure 2. Section view of a PW multilayer board.

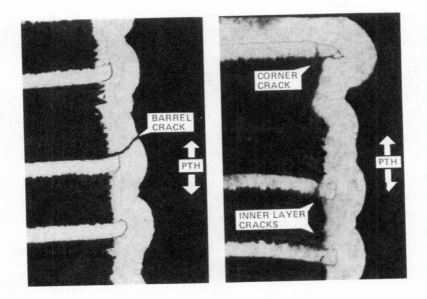

BARREL
CRACK

PTH

CORNER
CRACK

PTH

INNER LAYER
CRACKS

Figure 3. PWMLB plated-through-hole microstructures.

flexible circuits where consistently high ductility is essential. Obviously, Type W copper is not viable for PTH use, so in the case of multilayer circuits where rolled foil is used for inner layers, the PTH interconnections are still electrolytically deposited (Type E). At any rate, since Type W, Class 7 is a product of a totally different process, it makes a very good baseline for comparing metallurgical characteristics with the Type E classes.

Without too much difficulty, foil samples can also be obtained from an in-house PTH operation, in which case any of the procedures, standards, etc. developed for the commercial foils will apply to them so long as care is taken to avoid the introduction of plating artifacts that might influence the properties of the samples. This allows any of the controls that are used for commercial foil to be used for PTH copper, with the objective of gaining in-house PTH process control by this means.

The Control of Microcracking in Type E Copper

As had been mentioned earlier, the microcracking of copper interconnections has been found[3] to occur in high density PWMLBs regardless of the technology that is used in board fabrication. That is, cracks occur in inner layers or in the PTH barrel when any of the most advanced types of PWMLB structures are thermally stressed. There is evidence[4] that the stresses that cause microcracking are generated by the Z-direction (through-the-thickness) thermal expansion of the substrate material. The problem is that, when this thermal stress is applied to a Type E, Class 1 grade of copper, fracture without deformation will occur at low stress levels. The cracking phenomenon is thought[5] to arise from the combination of a thermal stress together with a copper plate in which grain boundary strength has been degraded. Further, this degradation is thought to be due to the presence of co-deposited impurities (esp. organics) that cause easy grain boundary separation at elevated temperatures. Some evidence of the lack of hot ductility in Class 1 copper can be seen from a comparison of the fracture surfaces of the foil samples in the scanning electron micrographs shown in Figure 4. In this case, the Type W, Class 7 material shows extensive shear and elongation in both the RT and the elevated temperature fractures. Although not as pronounced, the Type E, Class 3 material also exhibits a shear component at either test temperature. For Type E, Class 1 foil, the RT fracture shows slight deformation while the 550°F fracture is brittle.

When the ductile Class 3 and the brittle Class 1 foils are examined by metallographic sectioning, a similar comparison can be made, as is shown in Figure 5. The microstructure near the fracture exibits considerable elongation in the ductile foil and no elongation in the brittle material. This is an important distinction between the Class 1 and Class 3 types of material, and it is this difference that we must be able to detect if quality control is to be effective.

This also implies that Type E, Class 1 copper foil (or its PTH copper equivalent deposit) should not be used in PWMLBs where a Z-direction thermal stress is likely to be applied. When this condition exists, Type E, Class 3 (or its PTH equivalent) should be used. For foil, its quality can be controlled by proper procurement policy, and for a

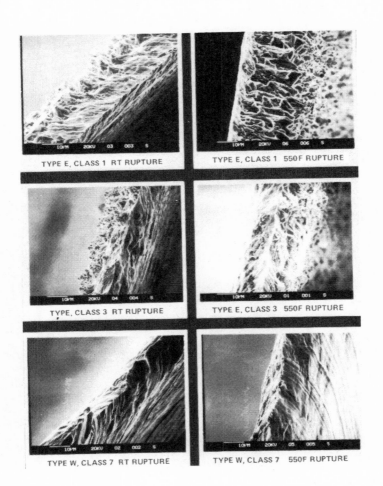

Figure 4. Copper foil rupture fracture surfaces.

plated-through-hole, quality can be maintained through the use of process controls that prevent the contamination of copper deposits, e.g., in a PWMLB PTH operation. One effective procedure for plating bath analysis resulted from the recent practical development of cyclic voltametry[6], and this technique is now available to the industry. However, along with a capability for monitoring the condition of the plating solution, there is a need for determining and monitoring the elevated temperature mechanical properties of the resultant deposit. Control over these properties can provide control of the microcracking phenomenon. In this paper, we suggest a test method for this purpose.

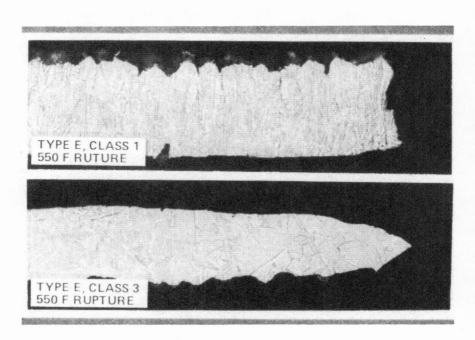

Figure 5. 550°F rupture microsections. Top: type E, class 1, brittle fracture; bottom: type E, class 3, ductile fracture 800X0.

Hot Rupture Testing

A procedure for the bulge testing of samples of electrolytic foil is described in some detail by Prater[7],[8] and Read as a result of their pioneering work in this area of technology. This contribution was followed by Lamb and co-workers[9] who determined the mechanical properties, including rupture pressure, of Type E copper as deposited from the commonly available copper plating solutions. More broadly, the Mullen hydraulic bulge test, or some adaptation of it, is widely used[10-11] for determining the stress-strain properties of sheet metals for the control of forming characteristics.

Rupture testing is uncomplicated, with no special requirement for sample preparation when testing thin sheets or foils. However, as our previous discussion indicates, determining the important properties for Type E copper in PWMLBs depends on having an elevated temperature testing capability that is not available on hydraulic machines. To overcome this

limit, a new method was established around the use of pneumatic pressure.
Figure 6 shows, in block form, our procedure for determining mechanical
properties by rupture testing. By this process, the effect of either a
dynamic or a static pressure can be determined at any temperature up to
550°F which is the highest processing, test or use temperature encountered
by most PWMLBs.

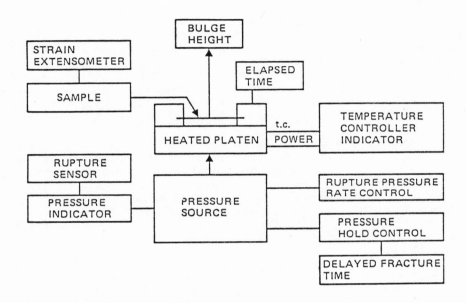

Figure 6. Procedure for the hot rupture testing of foil.

Dynamic Rupture Test. In this test, a constant loading rate of 1
psig/sec is applied to the test piece while it is held at a constant test
temperature. Measurements are made for bulge height at 10 psig intervals.
The bulge height at rupture and the rupture pressure for each test
temperature are recorded. Since this test yields data points for pressure,
bulge height and temperature, any combination of these variables can be
used for an evaluation of material properties. However, for quality
control purposes, a valid comparison between foil types is most easily
obtained by a plot of rupture pressure versus test temperature. From these
data, a measure of mechanical strength and ductility can be obtained.

When a pressure is applied to the foil, the foil deforms under
balanced biaxial tension, and this deformation continues until fracture
occurs. From the measured values for rupture pressure and bulge height,
from the geometry of an axisymmetrical bulge [1], and from membrane stress
and strain relationships [2] for thin walled pressure vessels, we can make
comparisons with values obtained from a uniaxial tensile test, as follows:

$$R = \frac{r^2 + h^2}{2h} \qquad [1]$$

where R is the radius of curvature, r is the radius of the aperture, and h is the height of the bulge. Then, we can obtain tensile strength from:

$$TS = \frac{PR}{2t} \qquad [2]$$

where TS is the nominal tensile strength, P is the rupture pressure, R is the radius of curvature and t is the original thickness. These mathematical relationships are more fully developed in the literature[12-15] where it is shown that obtaining true stress, true strain, nominal strain, ductility, and elongation values from bulge rupture tests can be achieved by taking into account the basic differences between the tensile test and the rupture test. However, this degree of detail is probably not necessary for effective use of the rupture test for quality control purposes.

Creep-Rupture Test. This procedure differs from the dynamic approach only in that a pre-set, constant pressure is applied and an up-counting timer is used for the measurement of the delayed-fracture time at constant pressure. Otherwise, the test equipment is identical. In either case, the assembly is designed such that tests can be made at any temperature between RT and 550°F, and at any pressure up to 80 psig. To allow the testing to be done on the same pressure scale, the cover plate aperture is variable for accommodating 1/2 ounce or 1 ounce foils.

Creep-rupture properties are obtained by the application of a constant pressure at constant temperature. Measurement is made for the time-to-rupture at each test pressure, and a plot of rupture pressure versus time at any test temperature will distinguish between the different types of copper. Typically, the nearer to the rupture pressure that the test pressure is set, and the higher the test temperature, the shorter is the time to fracture. Also, there is a distinctive difference in the delayed fracture characteristics of the various types and classes of foil and this allows their segregation according to rupture quality.

Mechanical Testing of Foil. Shown in Figure 7 is the mechanical set-up for the testing of 1/2 ounce and 1 ounce copper foil. For this test, sample pieces are cut into 4 X 4 inch squares. Nine test pieces are needed per lot, with three being used at each of three test temperatures (RT, 350, and 550°F) for a measure of dynamic rupture properties. For determining delayed-fracture characteristics, at least six test pieces are needed per lot, with three being used at each of two test pressures and one test temperature (350°F). As is true of the dynamic version, this test can be conducted over a much broader set of pressure, temperature or time test conditions if so desired.

From its flat initial condition, the resultant test piece takes on a hemispherical shape due to the applied pressure. The radius of curvature and the height of the bulge vary, of course, from sample to sample as can be seen from the collection of tested pieces shown in Figure 8.

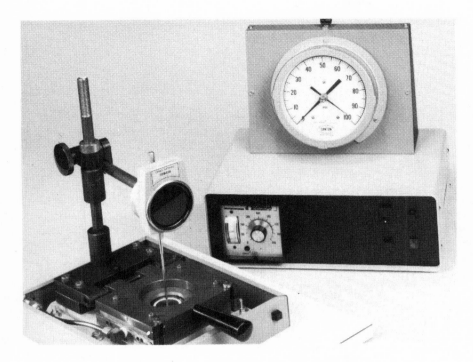

Figure 7. Foil rupture testing machine.

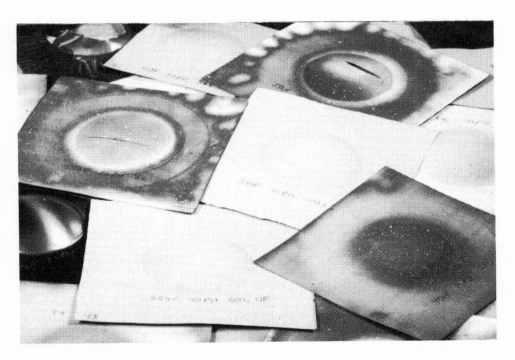

Figure 8. Typical rupture test samples: 10 cm square, aperture 2.

Setting Quality Control Standards – Dynamic Rupture

Numerous tests were made on 1/2 ounce and 1 ounce Type E and Type W copper foils. Commercial material representing foil production from the major foil suppliers was used to establish plots of rupture pressure versus test temperature for each type of foil at the two thickness levels. In an iterative process, the initial plots were used to classify larger numbers of foil and these results were then used further to refine the standards. Following this procedure, standard plots were developed and now exist for Type E, Class 1; Type E, Class 3; and Type W, Class 7 foils.

Figure 9 shows the effect of test temperature on the rupture pressure of 1 ounce Type E, Class 1 copper foil. Under these test conditions, the average rupture pressure at the solder-float temperature (550°F) is about 40% of that at room temperature. Of even more concern are the lowest rupture values, which have ranged down to 6 psig when the foil is stressed at the highest test temperature.

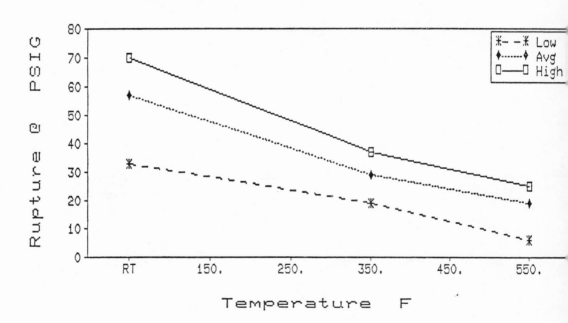

Figure 9. P versus T standard, type E, class 1, copper (1 oz).

This set of curves then are those that represent copper foil that is of Type E, Class 1 quality.

Figure 10 shows the effect of test temperature on the rupture pressure of 1 ounce Type E, Class 3 copper foil, and Figure 11 shows the results for Type W, Class 7 foil. These results show that the average rupture pressure at 550°F is over 60% of that at RT for these foils. Better yet, they maintain rupture strengths of over 35 psig at solder float temperatures, a significant improvement over Class 1 foil. The other observation worth mentioning is that the Class 1 foil properties extend

into the Class 3 and Class 7 property range at RT, so the elevated
temperature tests will provide a more positive indication of the correct
foil classification. In other words, RT testing is probably not sufficient
for proper foil characterization.

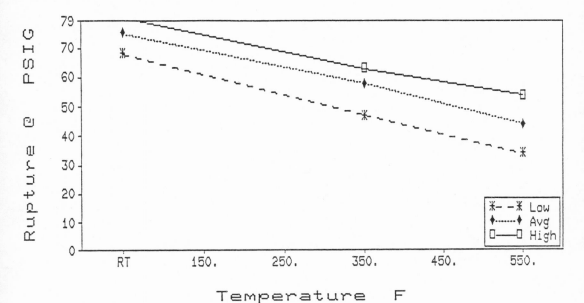

Figure 10. P versus T standard, type E, class 3 copper (1 oz).

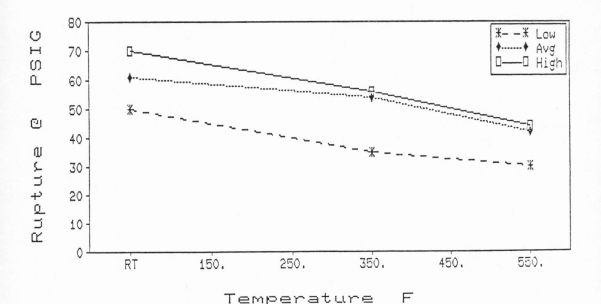

Figure 11. P versus T standard, type W, class 7 copper (1 oz).

Foil Characterization. Other than for material characteristics, rupture test results are influenced most by strain rate, by creep and by sample thickness. Shown in Figure 12 is the effect of strain rate for rolled, annealed foil. Figure 13 shows the effect of thickness for the various types of foils when a room temperature test is done at a strain rate of 1 psig/second.

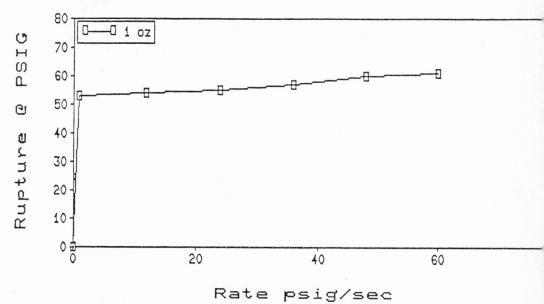

Figure 12. Effect of strain rate, type W, class 7 foil.

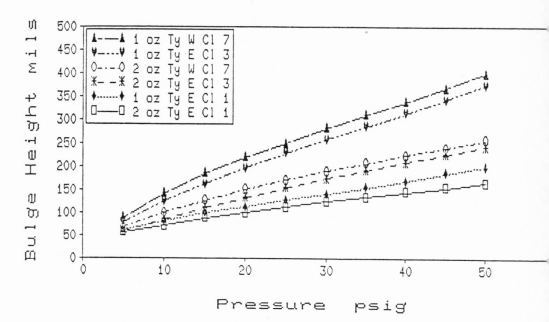

Figure 13. Effect of foil thickness, RT test, A2 Ty E and Ty W foils.

The standards that have been established for these three foils can now be used for the classification of foils from new lots of material, and this grading can be done before the new foil is laminated or used in PWMLB fabrication. The rupture properties for several different types and classes of foil are shown in Figure 14 where rupture pressure is plotted against the test temperature. These rupture-strength data were taken for 1 ounce foil that was obtained from commercial suppliers of electrolytic and rolled copper foil. As expected, when the different types and classes of foils are categorized by their comparison against the set of standard plots, each falls into the classifications which had been developed earlier.

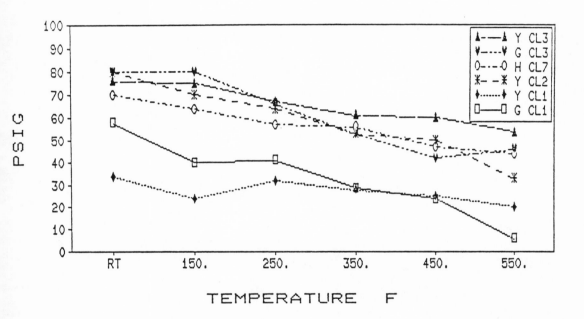

Figure 14. Rupture pressure versus temperature, Yates, Gould, and Hitachi foils (1 oz).

PTH Copper Characterization. As part of the classification process, foil samples from in-house PTH operations at several PWMLB facilities were made and tested. These samples represented copper deposits from the common plating solutions, including, sulfate, pyrophosphate and cyanide. When used in conjunction with the standards for the commercial foils, the PTH copper can be graded as being the equivalent of either the Class 1 or the Class 3 foil. In this way, PTH samples can be used to monitor the condition of the PWMLB plating process, i.e., through a determination of the mechanical properties at different times during the life of the plating bath.

As opposed to commercial foil, samples for the testing of PTH deposits must be obtained from the plating solution that is of interest. The objective is to make foil samples that can be used for the purpose of determining the PTH equivalent of foil properties. This will enable us to use the same standards for PTH deposits as were used for commercial foil. Shown in Figure 15 is the panel-loading end of an automated PWMLB plating operation. To make foil test pieces, a polished stainless-steel panel is loaded in the rack in preparation for copper plating. The panel will pass through this plating line where it will be coated with either 1/2 ounce o 1 ounce of copper. Although plating conditions for a large, flat panel ar not exactly like those in the local region of a PTH, the resultant deposi should be somewhat representative of the copper that would be deposited o a PWMLB. At the very least, foil made in this way provides a means for tracking the condition of a plating bath through the use of mechanical tests on the deposit, an approach that is commonly used for making tensil tests on plated deposits. Figure 16 shows how the plated copper is taken off the panel for testing. Once in foil form, the sample sheet can be sheared into 4 X 4 inch test pieces, to be tested for rupture strength in the same way that was previously described for the commercial foils.

Figure 15. PTH foil panel.   Figure 16. PTH foil sample.

In this manner, test pieces were made from a PWMLB production plating lin that uses acid copper sulfate plating chemicals. The initial bath

ke-up, with regard to composition, additives, etc. was done according to
e supplier's recommendations. A series of PTH foils were taken from this
ating operation over a period of eight months, and the foil was tested
r rupture properties by using the procedure described earlier. Shown in
gure 17 are the results of these tests.

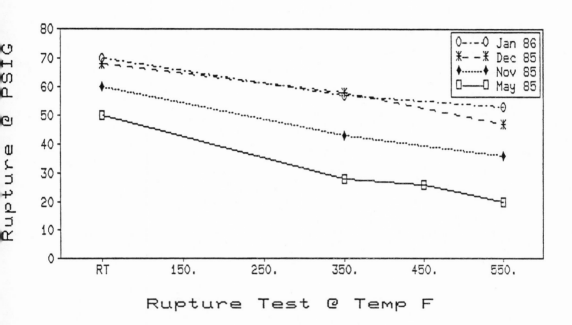

Figure 17. P versus T PTH Cu, acid copper (1 oz).

The easiest way to examine and to interpret these data is to overlay onto
his set of curves the standard plots for Class 1 and Class 3 foil. When this
s done, it can be seen that the May 85 PTH foil is in the average Class 1
ategory.   The November 1985 PTH foil is just above Class 1 and into the lower
evel of the Class 3 quality range. The December 1985 and the January 1986
TH foil materials are both well into the Class 3 quality range.

We mentioned that the plating bath represented by these foil samples
s an acid copper sulfate solution. It should also be mentioned that the
ath had been set up, in May 1985, to the supplier's recommendations. The
WMLB product from this plating line, at that time, exhibited PTH corner
racks in 50% of the boards. By November 1985, the plating bath was operating
ith modified quantities of additives, especially organic brighteners. This
hange is reflected in the improved rupture properties of the test pieces
s well as in the total elimination of PTH corner cracks.

Quality Control by Creep-Rupture Testing

When a foil test piece is held at constant pressure and constant
emperature, rupture occurs due to creep. A measure of time-to-fracture
rovides an indication of the creep-rupture resistance of the material

under test, and this characteristic varies according to the type of foil that is being tested. Using the static pressure test method, an evaluation was made of 15 lots of 1/2 ounce commercial foil. Shown in Figure 18 are the results as compiled for 10 lots of Type E, Class 1 foil.

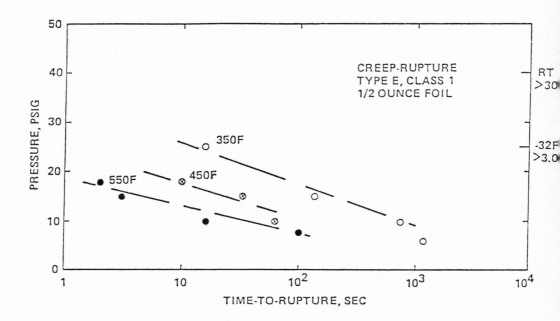

Figure 18. Effect of temperature on time-to-rupture at constant pressure type E, class 1 foil; 1/2 oz.

These test results show that the time-to-rupture increases with decreasing pressure and that the creep-rupture pressure decreases with increasing temperature. More importantly, for Class 1 foil, rupture will occur in less than 10 seconds at pressure levels of from 12 to 25 psig when the foil is stressed at temperatures above 350°F. This implies that Type E, Class 1 inner layer or PTH interconnections could be damaged during solder-float testing where the requirement is for a 10 second exposure at 550°F. Also, thermal cycling, soldering or solder repair could cause fracture with some very moderate Z-direction stresses.

Static rupture tests were made on material from 5 lots of 1/2 ounce Type E, Class 3 foil. The results are shown in Figure 19 where it can be seen that the same trends are present with the Class 3 material as were noted with the Class 1 foil. An important difference is that the Class 3 foil has much better creep-rupture strength at any test temperature, e.g., relatively high pressures are needed in order to cause fracture in times under 10 seconds. This difference in delayed-fracture characteristics makes it possible to distinguish a quality difference between the two classes of foil through the use of the static rupture test method.

Another phenomenon which is peculiar to Type E, Class 3 material is evident upon examination of the 550°F test results. Since early creep-rupture does not occur, the foil is annealed during the test cycle. This improves the rupture properties of the foil, as would an anneal before the rupture test. This does not happen with Class 1 foil during the 550°F test.

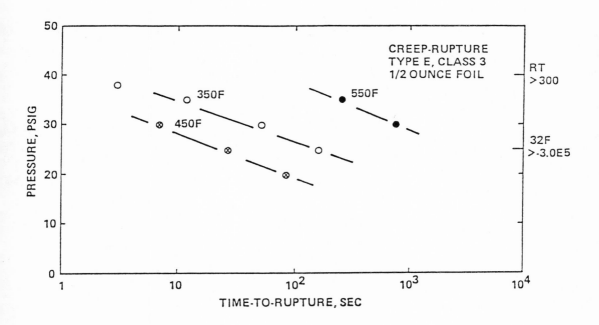

Figure 19. Effect of temperature on time-to-rupture at constant pressure, type E, class 3 foil; 1/2 oz.

Summary and Conclusions

PWMLB structures for high-speed, high-density circuits are prone to failure due to the microcracking of electrolytic copper interconnections. The failure can occur in the foil that makes up the inner layer traces or in the PTH deposit that forms the layer-to-layer interconnections.

In this paper, we show that there are some distinctive differences in the quality of Type E copper and that these differences can be detected before its use in a PWMLB. We suggest that the strength of some Type E copper can be very low when the material is hot and that it is the use of this poor quality material in a PWMLB that results in PTH and inner-layer microcracking.

Since the PWMLB failures in question are induced by a thermal stress, and since the poorer grades of Type E materials used in these structures are susceptible to premature failure under thermal stress, we propose the use of elevated temperature rupture and creep-rupture testing as a means for screening copper foil (or its PTH equivalent) in order to eliminate the problem of Type E copper microcracking in advanced PWMLBs.

References

1. R.A. Mogle and D.J. Sober, "Kevlar Epoxy for Chip Carrier MLBs", SI-928, NORTHCON '83, Portland, OR, May 11, 1983.

2. P. Hemler, et.al., "Hermetic Chip Carrier Compatible Printed Wiring Board", AFWAL-TR-85-4082, Final Report, July 1985, Contract No. F33615-82-C-5047, AFWAL/MLPO, WPAFB OH 45433.

3. G. Beene, et.al., "Manufacturing Technology for High Reliability Packaging Using Hermetic Chip Carriers (HCCs) on Compatible Printed Wiring Boards", Fifth Interim Report, June 1985, AFWAL Contract No. F33615-82-C-5071, AFWAL/MLTE, WPAFB OH 45433.

4. R.F. Clark, "Thermal-Mechanical Stress Analysis of Multilayer Printed Wiring Boards", GE Technical Information Series, R83ELS018, Nov. 1983.

5. R.F. Clark, H. Ladwig, and L. Zakraysek, "Microcracking in Electrolytic Copper", Printed Circuit World Convention III, IPC, May 1984.

6. D. Tench and C. Ogden, "Manufacturing Technology for PWB Electrodeposition Processes", AFWAL Contract No. F33615-81-C-5108, AFML, WPAFB, OH 45433, 1981.

7. T.A. Prater and H.J. Read, "The Strength and Ductility of Electrodeposited Metals, Part I", Plating, December 1949, p. 1221.

8. T.A. Prater and H.J. Read, "The Strength and Ductility of Electrodeposited Metals, Part II", Plating, Aug. 1950, p. 830.

9. V.A. Lamb, C.E. Johnson and D.R. Valentine, "Physical and Mechanical Properties of Electrodeposited Copper", J. Electrochemical Soc., Oct 1970, p. 341C.

10. F. Gologranc, "Automatic Determination of Stress-Strain Curves by Continuous Hydraulic Bulge Test", J. Mech. Eng., Strojniski Vestnik, Vol 26, No. 7-8, Ljubljana, July-Aug 1980.

11. R.F. Young, J.E. Bird and J.L. Duncan, "An Automated Hydraulic Bulge Tester", J. Applied Metalworking, ISSN 0162-9700/81/0701-0011, Vol 2, No. 1-11, American Society for Metals 1981.

12. H.M. Shang and T.C. Hsu, "Deformation and Curvatures in Sheet-Metal in the Bulge Test", J. Eng. for Industry, Aug. 1979, Vol 101, p. 341.

13. A.J. Ranta-Eskola, "Use of the Hydraulic Bulge Test in Biaxial Tensile Testing", Int. J. Mech. Sci., Vol 21, pp. 457-465, 1979.

14. A.S. Wifi, "Finite Element Correction Matrices in the Hydrostatic Bulging of a Circular Sheet", Int. J. Mech. Sci., Vol 24, No. 7, pp. 393-406, 1982.

15. M.F. Llahi, A. Parmar and P.B. Mellor, "Hydrostatic Bulging of a Circular Aluminium Diaphragm", Int. J. Mech. Sci., Vol 23, No. 4, pp. 221-227, 1981.

HETERODYNE HOLOGRAPHIC INTERFEROMETRY: HIGH-RESOLUTION RANGING AND DISPLACEMENT MEASUREMENT

James W. Wagner

The Johns Hopkins University
Center for Nondestructive Evaluation
Baltimore, Maryland

Heterodyne holographic interferometry (HHI) provides a means to map out-of-plane displacements and surface contours in full field with an out-of-plane resolution on the order of Angstroms. To gain an understanding of how this is done, three types of holographic interferometry will be considered - classical (homodyne) interferometry, quasi-heterodyne interferometry, and finally heterodyne interferometry. Note that each successive type represents an improvement, not only in resolution, but also in the dynamic range of the measurement.

Holographic Interferometry

Noncontact measurement of out-of-plane displacements over a range from:

- 125nm - 25um   (homodyne)

- 1.25nm - 25um (quasi-heterodyne)

- 0.125nm - 25um (heterodyne)

Relatively few applications of HI, Q-HHI, or HHI in electronics packaging or assembly are documented in the open literature. However, Q-HHI and HHI are relatively new techniques whose full potentials have yet to be realized.

Applications in Electronics

(documented)

- **Thermal-Mechanical Strain Analysis of Plated Through Holes in Multilayer Printed Circuit Boards**

- **Stress Analysis of Surface Mount Solder Connections**

- **Leak Detection and Measurement**

This list of potential applications comes from a greater familiarity with holographic measurement technology than with electronic packaging and assembly. Note that since these techniques actually measure optical path length differences, they could be applied to measure nonuniformities in the composition or stress distribution of optically transparent materials (including silicon in the infrared).

Other Applications

- **Warp and Flatness Measurements**

- **Vibration Analysis**

- **Uniformity of Properties of Materials for Integrated Optics**

- **Full Field Ultrasonic Detection for Debond and Composite Delamination**

Classical homodyne holographic recordings are made using a high-resolution photographic emulsion exposed to the interference of a modulated object beam with a reference optical beam. The fine interference pattern which is recorded may be used as a complex diffraction grating to reconstruct the object wavefront by illumination of the hologram with the reference beam. Such holographic plates may be doubly exposed so that a single reconstructing reference beam will reconstruct two object wavefronts. Differences between the two objects will then result in interference fringes superimposed on the viewed image. Object differences may be imposed by stresses applied between exposures. Alternatively, one may use optical "tricks" to create an effective surface displacement proportional to the contour or the surface.

Homodyne Construction

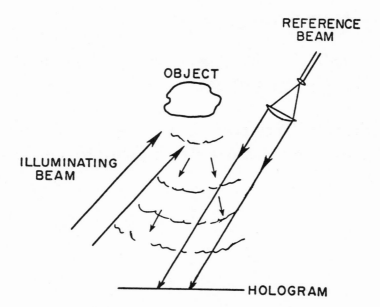

One "trick" for obtaining contour interferograms is to place the object to be contoured in a chamber filled with a fluid of known refractive index. Viewing the object through an optically flat window, an initial holographic recording is made. Prior to the second exposure, the fluid in the chamber is changed to one with a different refractive index. Upon reconstruction, the holographic interference pattern provides a contour map of the object surface.

In practice, we have used air as the refractive fluid in the test chamber. Its index is easily changed simply by changing the pressure within the chamber.

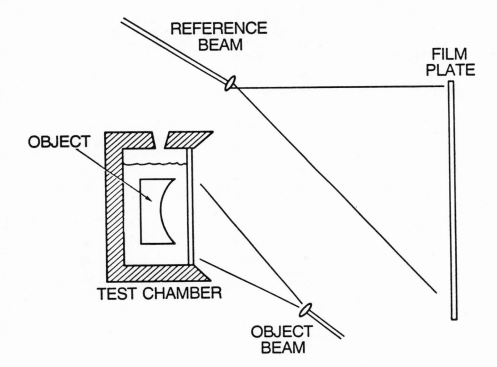

This display and the associated equation are keys to understanding the strength and shortcomings of holographic measurement techniques. The display is traced from an actual holographic contour interferogram. The contour of this object is somewhat like that of a spoon with its deepest point to the right of the picture. To interpret this homodyne interferogram, one could simply count the interference fringes. Obviously the resolution obtained by counting would be rather poor - perhaps half of a fringe, corresponding to an optical path length difference measurement resolution of one quarter of the optical wavelength (about .125 μm).

The actual image intensity as shown in the plot and expressed in the equation to the right is a <u>continuous</u> function of the surface displacement, D. Unfortunately, the intensity pattern is also a function of the average image intensity (without fringes) and the interferometric fringe contrast. Thus the homodyne interferogram intensity at any point may be represented as an equation in three unknowns where the desired variable, D, cannot yet be solved for directly.

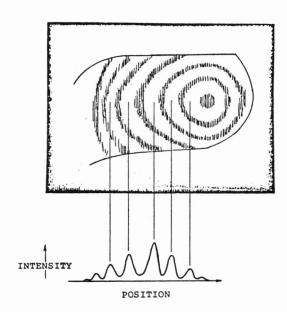

$$I = I_o\left[1 + C \sin\left(2kD\right)\right]$$

I = Measured Intensity

I_o = Image Intensity

C = Fringe Contrast

D = Displacement

INTENSITY

POSITION

Homodyne Display

Q-HHI and HHI offer two means by which one may solve for the variable, D. In order to do so, however, a new optical recording arrangement must be used. In this case, two angularly distinct reference beams are used sequentially in the double-exposure recording process - reference beam #1 for the first exposure then reference beam #2 for the second exposure after the object is stressed. Upon reconstruction reference #1 reconstructs the first object wavefront and reference #2 reconstructs the second.

Heterodyne Recording

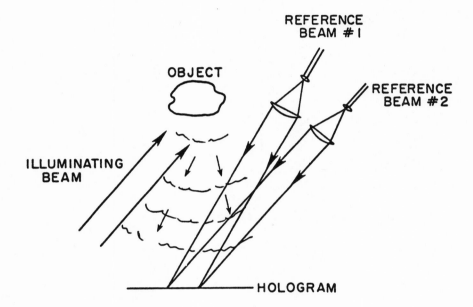

This figure illustrates the recording and playback setup which would be used to perform HHI contouring.

Contouring Using HHI

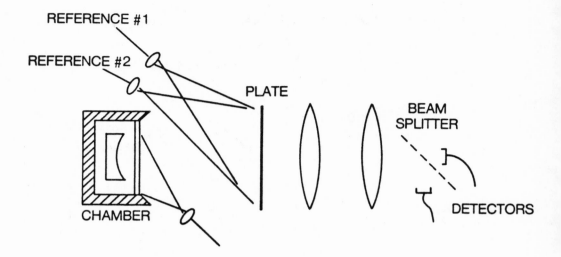

Once exposed and developed, the double exposure, dual reference beam hologram may be analyzed either by the quasi-heterodyne or heterodyne readout process. The Q-HHI process requires a digital image processing system to perform some arithmetic operations on the interference images. A phase shifter (optical delay) is placed in one of the reference beams and is used to shift the position of the fringes.

To perform the analysis, a homodyne image is viewed by the video camera and recorded in the digital image processor. Subsequently, a phase shift is imposed in one reference beam and a second image is recorded. After a second phase shift, a third image is recorded. At any point, the intensity of the image in the three video images may be represented respectively by the three equations shown. Note that these equations are essentially the same as the homodyne equations with the addition of a phase term contributed by the phase shifter. Since these phase terms are known, one is left with a system of three equations in three unknowns which can be solved directly for D.

Quasi-Heterodyne Readout

$$I_1 = I_0 \left[1 + C \sin(2kD + \Delta\phi_1) \right]$$

$$I_2 = I_0 \left[1 + C \sin(2kD + \Delta\phi_2) \right]$$

$$I_3 = I_0 \left[1 + C \sin(2kD + \Delta\phi_3) \right]$$

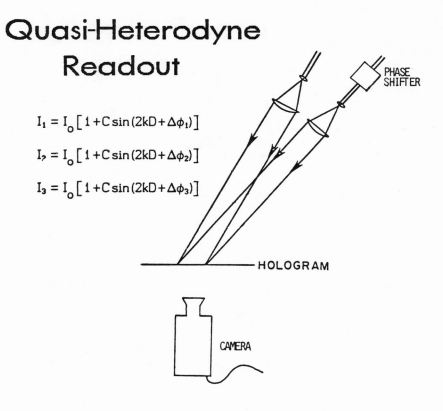

PHASE SHIFTER

HOLOGRAM

CAMERA

The result of the simultaneous solution involving the three images is itself an image whose pixel intensity is a function of elevation or displacement (left). Alternatively, the data may be displayed in perspective (right).

Wear Contours

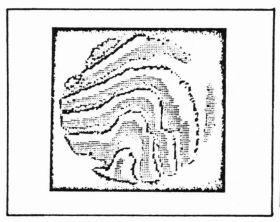 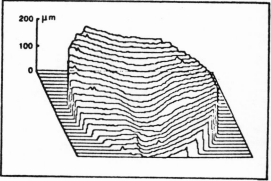

HHI readout requires that a frequency shifter be inserted into one of the reference beams. Typical shift values are 100 kHz and may be obtained with a combination of Bragg cell modulators. The resulting image intensity fluctuates sinusoidally with time at the shift frequency. While all image points vary in intensity at the same frequency, the <u>phase</u> at which the fluctuations take place is a function of local surface displacement or contour (D) as shown in the equation.

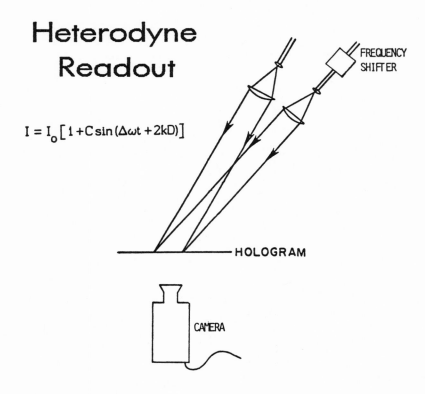

Heterodyne Readout

$$I = I_0 \left[1 + C \sin(\Delta \omega t + 2kD) \right]$$

FREQUENCY SHIFTER

HOLOGRAM

CAMERA

An image dissector camera or a translatable single-point detector is placed in the output plane of the interferometer. By comparing the phase of the intensity fluctuations at each point with the phase at some image reference point, the displacement, D, (or contour) may be directly measured independent of fringe contrast or spatial variations in object intensity. If an electronic phase meter is used whose accuracy is 0.360 degrees, for example, it is clear that one should be able to measure displacements to 1/1000 of a fringe! (At this point one begins to encounter some secondary spatial noise effects which may limit the certainty of such measurements.)

The fringes drawn in this figure are there for reference to the homodyne image. In heterodyning, the fringes "move" at the shift frequency and are not seen.

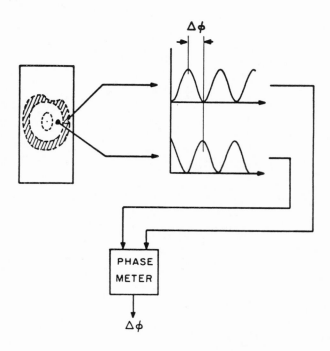

Contour map of a section of the spoon shaped object showing a wear gouge running down the center is shown. Note that the apparently poor resolution is the inability to "desensitize" holographic fringes by the double refractive index contouring technique. The actual optical path length difference resolution obtained during this experiment was about 3 Angstroms.

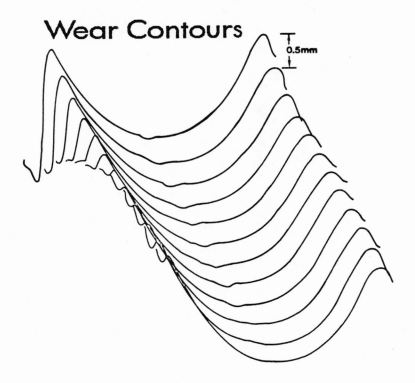

This HHI map of an acoustic surface wave propagating across an aluminum specimen demonstrates the ability to record and analyze high-speed, transient phenomena. The recording time was 9 nanoseconds. OPD resolution was 9 Angstroms.

Surface Acoustic Wavefront

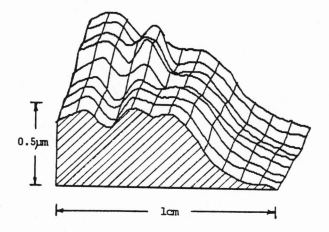

Several advantages are listed. Machine interpretability is unique to Q-HHI and HHI and holds promise for machine vision and intelligent processing (AI).

Advantages

Noncontacting

Very high out-of-plane resolution and dynamic range

Full field measurement and display

Machine interpretable

These limitations represent current state of the art in HHI and are <u>not</u> physical limits. Indeed they represent two of the challenges being addressed by the thrust of our current research activities.

Limitations

Highest resolution requires long analysis time and high readout system stability

Relatively poor (25-100μm) lateral resolution compared with out-of-plane resolution

"WHOLE WAFER" SCANNING ELECTRON MICROSCOPY

J. Devaney

HI-REL Laboratories
Monrovia, California

I. HISTORICAL PERSPECTIVE

1. **WHY SEM INSPECTION?**
    FAILURES DUE TO OPENS @ STEPS
    IN METALLIZATION.

2. **RESPONSE?**
    A PROLIFERATION OF SEM
    STEP COVERAGE SPECS STARTING
    WITH NASA GODDARD.

3. **ALTERNATIVES ?**
    STUDIES AT T.I. AND MOTOROLA
    FOR ALTERNATIVE SCREENS–

4. **RESULTS ?**

SEM INSPECTION REQUIREMENTS BECOME A MIL SPEC.

II. STEP COVERAGE PARAMETERS

1. OXIDE PROFILES

2. DEPOSITION TECHNIQUES

3. GLASSIFICATION TECHNIQUES
    INCLUDES DOPING

4. DIFFICULTY INCREASES AS
    NUMBER OF INTERCONNECT LAYERS
    INCREASES.

III. TODAY'S RESPONSE BY INDUSTRY

TEST CELLS

TEST DICE

COMPLEX DEPOSITION SYSTEMS

SEM SAMPLING

WHOLE WAFER INSPECTION 'ON LINE'

ALL THE MAJOR SEM MANUFACTURERS
ARE PROVIDING WHOLE WAFER SCANNING SYSTEMS

AMR–

CAMBRIDGE INSTRUMENTS–

ISI–

JEOL–

NANOMETRICS–

PHILLIPS ELECTRONICS–

IV. CHOICES/STATUS

PRESENT SPEC IS INADEQUATE

PRESENT SPEC IS OUT OF DATE

PRESENT SPEC IS IGNORED

PRESENT SPEC IS DIFFICULT

TO IMPLEMENT AND AUDIT

V. COURSES OF ACTION

CONTROL THE SYSTEM

CREATE – IMPLEMENT A NEW SPEC
WHICH RECOGNIZES THE NEED
FOR WHOLE WAFER SCANNING/INSPECTION

IGNORE THE SYSTEM

VI. WHOLE WAFER SCANNING EXPOSES THE WAFER TO:

HANDLING

A VACUUM: CLEAN - DIRTY

ELECTRON BEAM

WAFER HANDLING SYSTEMS HAVE MATURED

TO THE POINT THEY ARE

AN OFF-THE-SHELF ITEM

VACUUM SYSTEM MUST BE MAINTAINED

AND PROVEN NOT TO CONTAMINATE SAMPLES

DAMAGE OF DEVICES BY EXPOSURE

TO AN ELECTRON BEAM IS STILL

NOT COMPLETELY UNDERSTOOD

– LOW VOLTAGES,   < 2KV ARE PROBABLY
                                      REQUIRED

A LARGE PERCENTAGE OF DEVICES STILL FAIL

THE SEM STEP COVERAGE REQUIREMENTS

THIS WILL WORSEN AS TWO

AND

THREE LEVEL INTERCONNECT SYSTEMS PROLIFERATE

AN INTERACTIVE GROUP OF INDUSTRY AND GOVERNMENT SHOULD GENERATE A SET OF GUIDELINES OR SPECIFICATIONS GOVERNING WHOLE WAFER INSPECTION

HANDBOOK OF CONTAMINATION CONTROL IN MICROELECTRONICS
Principles, Applications and Technology

Edited by

Donald L. Tolliver
Motorola, Inc.

Contamination control technology is now a prerequisite of modern electronics. This has not always been the case. However, since about 1980, advanced microelectronic circuitry has increased dramatically in its complexity and degree of integration or density of active components, thus necessitating meticulous contamination control. The one megabit DRAM is in production; the 4 megabit device is forecast for 1990; 16 and 64 megabit capacities are in the planning stages; and the Japanese envision a 100 megabit device. Obviously, device defect density is or will be so critical to the successful manufacturing of these devices that only the most astute companies with advanced contamination control technology will be able to survive in the marketplace. For this very basic reason, this handbook will have a timely and important role to play in the industrial marketplace.

This book introduces contamination control in a relatively comprehensive manner. It covers the basics in most areas for the beginner, and it delves in depth into the more critical issues of process engineering and circuit manufacturing for the more advanced reader. The reader will begin to see how the puzzle of contamination control comes together and to focus on the fundamentals required for excellence in modern semiconductor manufacturing.

What makes the arena of contamination control unique is its ubiquitous nature, across all facets of semiconductor manufacturing. Clean room technology, well recognized as a fundamental requirement in modern day circuit manufacturing, barely scratches the surface in total contamination control. This handbook makes the first attempt to define and describe most of the major categories in current contamination control technology.

CONTENTS

ISBN 0-8155-1151-5 (1988)

488 pages

SEMICONDUCTOR MATERIALS AND PROCESS TECHNOLOGY HANDBOOK
For Very Large Scale Integration (VLSI) and Ultra Large Scale Integration (ULSI)

Edited by

Gary E. McGuire
Microelectronics Center of North Carolina

This handbook is a broad review of semiconductor materials and process technology, with emphasis on very large scale integration (VLSI) and ultra large scale integration (ULSI). The technology of integrated circuit (IC) processing is expanding so rapidly that it can be difficult for the scientist working in one area to keep abreast of developments in other areas of the field. This handbook solves this problem by bringing together "snapshots" of the various aspects of the technology.

The generally accepted goals of VLSI are to produce devices with 100K gates, or memory bits, per circuit, or devices with geometries of less than 1 μm. The goals of VLSI technology are a natural evolution of current IC technology with the horizontal dimensions, gate length, junction depth and oxide thickness decreasing with each succeeding generation of product.

The methodology necessary for achieving the goals of VLSI is an extension of available lithographies used for pattern definition. However, new process technologies have been developed to minimize process-induced degradation of the defined geometry. In addition, new materials have been developed to meet the increasing demands that result from the decreasing geometries.

Many of these developments have never before been reviewed in a formal text, especially in one that discusses all major aspects of device processing. Several areas discussed in the handbook, such as orientation dependent etching, have remained trade secrets until just the last few years. There is also a section on packaging which is an area which has often been neglected but which is becoming an increasingly complex technology.

CONTENTS

ISBN 0-8155-1150-7 (1988)

675 pages

RADIATION EFFECTS ON AND DOSE ENHANCEMENT OF ELECTRONIC MATERIALS

by

J.R. Srour
Northrop Corporation

D.M. Long et al
Science Applications, Inc.

This book describes radiation effects on and dose enhancement factors for electronic materials.

Alteration of the electrical properties of solid-state devices and integrated circuits by impinging radiation is well-known. Such changes may cause an electronic subsystem to fail, thus there is currently great interest in devising methods for avoiding radiation-induced degradation. The development of radiation-hardened devices and circuits is an exciting approach to solving this problem for many applications, since it could minimize the need for shielding or other system hardening techniques.

Part I describes the basic mechanisms of radiation effects on electronic materials, devices, and integrated circuits. Radiation effects in bulk silicon and in silicon devices are treated. Ionizing radiation effects in silicon dioxide films and silicon MOS devices are discussed. Single event phenomena are considered. Key literature references and a bibliography are provided.

Part II provides tabulations of dose enhancement factors for electronic devices in x-ray and gamma-ray environments. The data are applicable to a wide range of semiconductor devices and selected types of capacitors. Radiation environments discussed find application in system design and in radiation test facilities.

CONTENTS

ISBN 0-8155-1007-1 (1984)

128 pages